TRUTH
&Assertibility

TRUTH &
Assertibility

Nik Weaver

Washington University in St. Louis, USA

 World Scientific

NEW JERSEY • LONDON • SINGAPORE • BEIJING • SHANGHAI • HONG KONG • TAIPEI • CHENNAI

Published by

World Scientific Publishing Co. Pte. Ltd.

5 Toh Tuck Link, Singapore 596224

USA office: 27 Warren Street, Suite 401-402, Hackensack, NJ 07601

UK office: 57 Shelton Street, Covent Garden, London WC2H 9HE

Library of Congress Cataloging-in-Publication Data
Weaver, Nik.
   Truth and assertibility / by Nik Weaver (Washington University in St. Louis, USA).
     pages cm
   Includes bibliographical references and index.
   ISBN 978-9814619950 (hardcover : alk. paper)
   1. Frege, Gottlob, 1848–1925. 2. Tarski, Alfred. 3. Arithmetic--Foundations. 4. Logic, Symbolic
and mathematical. 5. Mathematics. I. Title.
   QA9.W383 2015
   510.1--dc23
                          2015007707

British Library Cataloguing-in-Publication Data
A catalogue record for this book is available from the British Library.

Printed in Singapore

On this view there are no philosophical problems with truth; all the real philosophical problems concern the notion of *assertibility*.

Hilary Putnam, "On Truth"

Preface

In the philosophy of mathematics there are two rival conceptions of truth, a classical one and a constructive one. Classically, truth is thought of as relating language to the world: to say that a sentence is true is to say something about how it compares to some objective reality. The constructive interpretation drops the overt reference to external reality and identifies truth with provability — in the informal sense of having a conclusive rational demonstration, not the syntactic sense of being formally derivable in some axiomatic system. Thus, a sentence is constructively "true" if we have an unimpeachable right to assert it.

More neutrally, we may borrow a term from the philosophy of language and call the constructive notion *assertibility*. I imagine most classical mathematicians would probably suppose either that this notion is effectively captured by the syntactic characterization of derivability in first-order predicate calculi, or if not, that it is too imprecise to support formal analysis. My main aim is to show otherwise: the notion of assertibility cannot be reduced to syntactic derivability, but nonetheless it is meaningful, interesting, amenable to formal analysis, and what is more, essential for the resolution of a variety of philosophical problems that are classically intractable.

The core novelty of this book is an axiomatization of the concept of assertibility (Section 4.5). Despite the centrality of assertibility to constructive mathematics, there has been little previous work in this direction. The likely reason is that most constructivists apparently consider the statement that a given sentence is assertible to simply be equivalent to that sentence, which trivializes the problem.

As plausible as this equivalence may seem (on a constructive interpretation of "equivalence"), I claim a careful analysis of its apparent justification reveals that one direction only works as a deduction rule, not as a direct implication (Sections 4.4 and 4.5). This is *good* because the full statement of equivalence is inconsistent. There is a paradox involving a sentence which denies its own assertibility, for instance.

In my axiomatization such paradoxes cannot be derived. Indeed, the axioms for assertibility are provably consistent both on their own (Section 5.1) and in conjunction with elementary reasoning about either numbers (Section 5.2) or concepts (Sections 5.5 and 5.6). The number-theoretic assertibility system has the remarkable additional property that it effectively derives its own soundness and consistency (Theorem 5.2). It is the distinction between a sentence and its assertibility that gets us past the apparent obstacle of Gödel's incompleteness theorem at this point.

I want to emphasize that the paradoxes are obstructed for principled reasons, not by some artificial device. It is also noteworthy that there are compelling reasons for using intuitionistic logic when reasoning about assertibility, and indeed intuitionistic logic is also essential to the consistency results. Often, switching between classical and intuitionistic logic does not fundamentally alter the deductive strength of a formal system, but it has a dramatic effect here: switching to classical logic would allow us to prove that a falsehood is assertible.

There is no comparable resolution of the classical paradoxes about truth. In this setting attention has turned to the technical problem of constructing classical truth predicates for various classes of sentences and with various properties, and there is a substantial literature on this subject. However, one faces a basic conceptual difficulty in saying exactly what constitutes a classical truth predicate. Tarski's solution to this problem, while ingenious, fails to capture essential aspects of our intuitive notion of classical truth. Surprisingly, as Putnam realized ([72], p. 320, the source of my epigraph), this difficulty can be resolved using assertibility (Section 4.6). This shows that assertibility is conceptually prior to the global notion of a classical truth predicate. I hasten to add that one does not need assertibility to construct classical truth predicates or to understand their behavior in any particular case.

Another fundamental problem that cannot be solved classically concerns the inconsistency of Frege's formal theory of concepts. As emphasized by Quine, the underlying error that leads to this inconsistency is a use/mention confusion. However, this observation alone cannot be considered a complete solution because it leaves us with no meaningful ability to quantify over concepts. It is intuitively clear that there are general truths about concepts; there must be some way of expressing them. The solution is to use an assertibility predicate where one would classically place a truth predicate (Sections 5.5 and 5.6).

I came to this subject as a classical mathematician, and like most mathematicians, probably, I was initially hostile to the psychologistic tinge of constructivism. But I have come to appreciate that the classical conception of truth as expressing a correspondence between a language and some external reality breaks down when the language is itself part of the reality it is being used to describe. Nor do I see anything particularly mysterious about this failure. What I find remarkable is that the constructive approach gives us a way to make sense of this kind of self-referential language. This means that intuitionistic logic is bound to be an essential tool for reasoning globally about mathematical reality, since there is no external perspective from which to view all of reality, which means that any such reasoning is likely to have a circular aspect. In short, I have a route to intuitionistic logic that I think is essentially forced.

A word about terminology. Since I am interested in both classical and constructive truth, and since I think the classical version has at least historical priority, I felt it made sense to use the word *truth* exclusively to refer to classical truth. I use *assertibility* for constructive truth, i.e., provability in the informal semantic sense. By *proof* I will mean a semantically valid argument and by *derivation* a formal deduction that is executed within some axiomatic system.

Nik Weaver

Contents

Preface   vii

1. Truth   1

    1.1   The liar paradox   1
    1.2   Sentences and propositions   3
    1.3   The T-scheme   8
    1.4   Numerical truth   15
    1.5   Arithmetic   22
    1.6   Arithmetic mod n   26
    1.7   Truth and meaning   27

2. Concepts   31

    2.1   Predicates and concepts   31
    2.2   Russell's paradox   34
    2.3   Interpreted languages   38
    2.4   Global truth   42
    2.5   Defining truth   47
    2.6   The revenge problem   54
    2.7   Second-order logic   58

3. Deduction   61

    3.1   Natural deduction   61
    3.2   The completeness theorem   66
    3.3   Peano arithmetic   70
    3.4   The first incompleteness theorem   72
    3.5   The second incompleteness theorem   76
    3.6   Arithmetic with truth   79

3.7   Formal truth predicates   83

4.   Assertibility   89

4.1   Assertibility as a concept   89
4.2   Proofs and meaning   94
4.3   Existence   99
4.4 A versus $\mathbb{A}(A)$ 104
4.5   Axiomatizing assertibility   110
4.6   Applications   117

5.   Systems   123

5.1   Pure assertibility   123
5.2   Arithmetical assertibility   127
5.3   Problems of rational agency   132
5.4   Weak interpretation   136
5.5   Conceptual assertibility   143
5.6   Second-order logic   149

6.   Surveyability   151

6.1   The iterative conception   151
6.2   Indefinite extensibility   155
6.3   Surveyable concepts   159
6.4   Abstract objects   164
6.5   Conclusions   168

Notes   175

Bibliography   181

Notation Index   187

Subject Index   189

Chapter 1

Truth

1.1 The liar paradox

The liar paradox. Excluded middle not essential. Quine's version. The significance of the problem. Two camps.

We are going to investigate the idea of truth as it arises in mathematics and logic. The liar paradox is a good place to start, since it represents a basic difficulty we have to face.

Consider the *liar sentence*

This sentence is not true.

Evidently, if it is true then it is not true, and if it is not true then it is true. This is the *liar paradox*.

The first impression most people have about the liar sentence is that it is completely meaningless. But surely a meaningless sentence cannot also be true. So if it is meaningless then in particular it is not true, and that is just what it says of itself ... which would make it true. We have reached a contradiction again.

The argument I gave above has the form "if the liar sentence is true, then X, and if it is not true then Y". Maybe the problem is this dichotomy. Should we question the implicit assumption that the liar sentence is definitely either true or not true (an example of the sometimes controversial *law of excluded middle*)? Surprisingly, it does not matter; this assumption is not essential to the paradox. We can make the following alternative argument:

1. If the liar sentence were true then it would not be true, a contradiction.

2. Step 1 shows that assuming the liar sentence is true leads to a contradiction. So it cannot be true.
3. But that is just what it says, so it must be true. Contradiction!

This version of the argument does not initially take it for granted that there is a definite answer to the question of whether the liar sentence is true. It simply observes that there is no way it can be true, since that would lead to a contradiction. Therefore it must not be true, and that is enough to get the vicious cycle going.

Another idea is that self-reference may be the illegitimate element. If this is the real source of the paradox, then the solution might be to strike the phrase "this sentence" from our language. That would prevent us from being able to even formulate the liar sentence and would thereby seemingly nip the problem in the bud. But even this procrustean solution fails: Quine has formulated a version of the liar sentence which escapes it.[1] Consider the sentence

"Yields falsehood when appended to its quotation" yields falsehood when appended to its quotation.

A little thought shows that this sentence asserts of itself that it is false, similarly to the standard liar sentence. But it does so in a roundabout way, without mentioning itself directly. The moral is that as long as we have a basic ability to manipulate syntax, we can design a sentence that refers to itself indirectly, by means of a syntactic construction that is rigged to generate the sentence itself. So self-reference really cannot be avoided.

The liar paradox does not have a trivial resolution. This is puzzling because the paradoxical derivation is so simple that it seems like the flaw should not be that difficult to spot. Even worse, the paradox is not only resistant to easy solutions, it is resistant to hard solutions too: many sophisticated theories about truth have been proposed as ways of getting us out of the difficulty, but they always seem to have the dismaying effect of generating more complicated paradoxes. (This is the so-called "revenge problem", and I will say a bit more about it later.)

At this point one might question whether it is worth putting any further effort into the issue. The whole thing seems rather frivolous. Does anything important hinge on what we decide to say about the liar sentence?

I agree that the liar sentence is not so important in itself. But what concerns me is that our inability to resolve the paradox seems to reveal some basic failure of understanding, either about the nature of truth or about elementary logical reasoning. That failure of understanding is what matters. We ought to figure out what is going on, not only because, without knowing where the problem is, we have no way to tell what kind of repercussions fixing it might have — but also, as Hilbert said in a different context, for the sake of "the dignity of the human intellect itself".[2]

People who work on the liar paradox seem to fall into basically two camps. One camp sees it as fundamentally a technical problem. On this view, there is nothing mysterious about language. We create it and it behaves the way we specify it to behave. All the liar paradox shows is that if we include the word "true" in our language and declare that it is to function in a certain way, then we are going to get contradictions. Perhaps this is surprising because the same word with the same qualities can be used in a variety of ways without causing any trouble. So the appropriate response is, first, to make all the properties we intuitively attribute to the word "true" explicit, and then to cleverly weaken one or more of them in such a way as to block the liar paradox but leave intact the basic functionality of the word. There need not be any unique correct solution; there might only be a variety of technical schemes, each of which has its own merits and drawbacks.

The opposing attitude considers truth to be a real, well-defined quality that exists independently of us, in some sense. Language is not just a formal game; words have meanings, including the word "true". From this standpoint the lesson of the liar paradox is that we have somehow gotten truth wrong. Either it lacks some property we think it has, or it cannot be reasoned about in some way that we think it can, or there is just some aspect of it that we have to understand better. The problem is therefore primarily conceptual, though its solution may involve technical work.

I think both views are partially right, but maybe I will not try to explain what I mean by that quite yet.

1.2   Sentences and propositions

> *Realism versus nominalism. Can a string of symbols be true? Properties of which symbolic strings are the primary bearers. Sentences not referential. Propositions as meanings. Propositions as referents of that-clauses. The liar paradox for propositions.*

Before we go any further, we have to address a slightly distracting side issue. Many philosophers consider truth to be fundamentally an attribute not of sentences but of some more abstract correlate of sentences called "propositions". The idea is that sentences function by referring to or expressing abstract propositions, and it is these propositions which are the "primary bearers of truth". This seems to be a common opinion, but it is controversial, with some dissenters denying that there even are such things as propositions. Let us try to briefly clarify what is at issue.

The question of whether propositions exist is an example of a realist/nominalist debate in which the *realist* affirms the literal existence of abstract entities of some kind and the *nominalist* denies this. In such debates, both sides typically carry heavy burdens. The realist has to explain what it even means to say that "abstract entities" of some kind exist — in Wittgenstein's famous phrase, what would it be like if they did not? — and assuming they did, how we could know anything about them. The nominalist, on the other hand, has to face the brute fact that we constantly, and apparently cogently, talk about abstract entities as if they were real things. If such talk is not simply gibberish, then there must be a way to make sense of it, and the nominalist owes us an explanation of how to do this. Ideally, we would want a systematic method for rephrasing away the offending terms.

For instance, syntactic objects like sentences are already somewhat abstract. (The "same" sentence can appear in different settings — in two different books, say — showing that sentences are not physical objects, at least not of a conventional kind.) But they do not seem particularly mysterious or occult, and it does not seem like it should be that hard, in principle, to eliminate most of the things we say about abstract sentences in favor of more complicated statements about physical sentence "tokens" (the actual printed words in a book, and so on).

I will have a little more to say about the realism versus nominalism issue in Chapter 6, but I will mostly try to avoid it. I plan to do this by generally using realist language without comment, but with occasional hints about how it might be nominalistically reinterpreted.

Now let us turn to the question of whether truth is more sensibly seen as a property of sentences or propositions. One motivation for the propositions-as-primary-truth-bearers view could be an intuitive sense that sentences are the wrong kinds of things to be true or false. A sentence is merely a string of symbols. So saying that the sentence "snow is white" is true might sound a little like saying that the numeral "37" is prime. Surely what is really prime is not the two-digit string "37" but the abstract thing that string represents, namely, the actual number 37. Similarly, the reasoning might go, what is true must be not the literal sentence "snow is white" but the abstract proposition which that sentence represents or expresses.[3] As Frege said, "when we call a sentence true we really mean its sense is".[4]

It certainly seems implausible that snow being white, or it being true that snow is white, should have much to do with any intrinsic property of the symbolic string "snow is white". But the analogy I just made is unfair. The numeral "37" fails to be prime for the uninteresting reason that the concept of primality by definition applies to numbers and not to numerals. The right property to consider here is the property not of being a prime number but rather of being a numeral which represents a prime number. And "37" does indeed, uncontroversially, have this property.

Of course, being the numeral representation of a prime number is defined in terms of on the abstract concept of primality, so it might seem reasonable to say that abstract numbers are the "primary bearers" of primality, and the property of being the numeral representation of a prime number is merely derivative.

So it might seem, but let us put this question on hold and take a moment to ask, of what sort of properties would symbolic strings legitimately qualify as the primary bearers? One might propose properties along the lines of being composed entirely of lowercase consonants, say, or not containing the string "abc", as playing this role. These are properties that symbolic strings can have on their own merits, as it were. Generalizing, consider any computer program that has the following structure: first, it collects a symbolic string as input; next, it performs some sort of computation using that string; finally, it returns either a YES or NO output, depending on the results of the preceding computation. Given such a program P, we could call a symbolic string P-*accepted* if it elicits a YES output and P-*rejected* if it elicits a NO output.

Readers with a computer science background should not have any trouble convincing themselves that a program could be written that would accept those and only those strings that are composed entirely of lowercase consonants, or those and only those strings that do not contain the string "abc". Obviously, many other properties of symbolic strings could be framed in terms of P-acceptance for an appropriate program P. The claim I want to make now is that any such property could be considered a property of which symbolic strings are the primary bearers, to the extent that the notion of being a "primary bearer" of a property is even cogent. Is this overgeneralizing? It is hard to see how, since any program is intrinsically nothing more than a series of instructions for manipulating symbolic data. We feed the program a string; it perfoms a computation and gets a result. Inducing a certain program to produce a certain result is a property that some strings enjoy and others do not. In some cases it might be possible to explain the string having this property in terms of more abstract concepts, but that does not make it any less directly a property of the string.

If the claim is granted, this answers the question about primality in an unexpected direction. It is certainly possible to write a program that collects a string of symbols as input and returns a YES output if the string is the numeral representation of a prime number, and a NO output if it is not. (I will leave it to readers who are interested in programming to figure out how.) From what I said above, it follows that being the numeral representation of a prime number is a property of which symbolic strings are the primary bearers. It need not be considered secondary to some other more abstract property.

This suggests that the intuition that syntactic objects cannot have substantial properties, or can only do so in some derivative fashion, is mistaken. So the idea that a syntactic object like a sentence could not possibly bear a deep property like truth is not obviously right. In fact, we will see in Sections 1.4 through 1.6 that there are settings in which truth looks very similar to P-acceptance for a suitable program P, and I will return to the propositions-as-primary-truth-bearers question then.

A second comment about the "37" analogy is that there is an important grammatical difference between the two cases. The numeral "37" functions as a noun, and it is the job of nouns, or more generally noun phrases, to refer to objects. So it at least makes grammatical sense to say that the numeral "37" refers to the abstract number 37, as it makes sense to say that the phrase "the Eiffel Tower" refers to the iron lattice tower located on the Champ de Mars in Paris, whereas it does not make sense to say that the sentence "snow is white" refers to some abstract proposition, because "snow is white" is not a noun phrase. This may seem obvious, but it needs to be said because there is a tradition in the philosophy of language, going back to Frege and Russell, which considers sentences to *refer to* truth values or propositions in the same way that nouns refer to objects.

One way to make it clear that sentences do not perform a referential function is by observing that inserting a sentence into a noun phrase position typically results in grammatical nonsense. " 'Snow is white' is a sentence" is true, but "snow is white is a sentence" is nonsensical. This simple observation shows that sentences do not have referents. (Incidentally, we also see here that enclosing a sentence in quotation marks turns it into a name for that sentence, which then functions as a noun.)

This comment about grammar is relevant to our current discussion because we would like to figure out what kind of things propositions could be, and it shows that *referents of sentences* is not a possible answer, because there are no such things.

However, there are other potential answers. For instance, we might want to say that propositions are the *meanings* of sentences. This ties into a second motivation for considering propositions, not sentences, to be the primary bearers of truth: a single sentence can change its meaning depending on context, so that it is sometimes true and sometimes false. ("It is raining here now.") It is natural to conclude that the meaning of the sentence must be the thing that is actually true or false.

On the other hand, this suggestion takes "meanings" to be some sort of independent objects, and one is naturally skeptical of such a move. For instance, it is not obvious that there is even a clear criterion which characterizes when two sentences have the same meaning, which would seem to be a precondition for taking the meanings-as-things theory seriously. In practice, we tend to say that two sentences have the same meaning if they are easily seen to imply each other, but not if the equivalence is unobvious, and there is no sharp dividing line between the two cases. Thus, it seems more straightforward to simply conclude that the truth of a sentence can change depending on context or interpretation, rather than that truth is really an attribute of some separate entity, the sentence's meaning.

A different defense of propositions could be made on grammatical grounds. Although sentences are not noun phrases and hence cannot refer, every sentence can be nominalized into a "that-clause" (e.g., *that snow is white*) which does exhibit the right form to have a referential function. It seems natural to take propositions to be the referents of such clauses. We can use the word "expresses" to describe the relation between the original sentence and the corresponding proposition; thus "snow is white" expresses the proposition that snow is white.

That such clauses can function as noun phrases is clear; this sentence is an example. How would a nominalist account for them? This might be possible by an alternative parsing that groups the word "that" with the verb rather than the embedded sentence. Thus, instead of reading "knows" as a verb that combines with the two noun phrases "John" and "that snow is white" to make a sentence, we can read "knows that" as a pseudo-verb that combines with a noun phrase, "John", and a sentence, "snow is white", to make a sentence. That-clauses might thereby be explained away as grammatical illusions.

In any case, propositional attitudes like knowing, believing, doubting, and so on, raise many interesting questions, but this line of thought takes us into the philosophy of language and away from the main focus of this book, which is the universal, factual statements of mathematics and logic. For our purposes the appropriate criterion of identity for propositions is: two sentences express the same proposition if each implies the other. So I will simply take this as a definition. However, there is a subtlety here that I am not prepared to discuss yet, so I am going to work at the level of sentences for the rest of this chapter and come back to the subject of propositions in Chapter 2. Meanwhile, I believe that the sense of incredulity that sentences could possibly bear a property like truth is a confusion that can be dispelled, and I hope to do so later in this chapter.

Before leaving this topic let me mention that (as many have noted) the liar paradox does not seem particularly sensitive to the sentence/proposition distinction. The liar sentence can be reformulated at the level of propositions like so:

> This sentence does not express a true proposition.

Notice that this version is able to refute the diagnosis that it fails to express any proposition. For if it expresses no proposition, then in particular it does not express a true proposition, which is just what it says. So this would show that it does in fact express a proposition, indeed a true one. And that leads to a contradiction in the usual way.

1.3   The T-scheme

Tarski's T-scheme. Ordinary versus schematic variables. The T-scheme as a correspondence theory. Indirect assertions. Disquotation and use versus mention. The Tarskian catastrophe. The liar paradox revisited. Truth predicates.

What does it mean to affirm that a sentence (or proposition) is true? "Snow is white" says something about snow, while " 'snow is white' is true" says something about the sentence "snow is white". But what exactly?

Saying that a sentence is true seems to indicate some kind of correspondence between the sentence and external reality. "Snow is white" is true because out in the world, snow really is white. Tarski was able to turn this intuitive idea into a formal condition.[5]

His informal thinking goes like this.[6] Whenever we are willing to affirm that snow is white, we should also be willing to affirm that "snow is white" is true, and conversely. We would never find ourselves in a situation where we agreed that snow is white but not that the sentence "snow is white" is true, nor would we ever agree that "snow is white" is true but not that snow is white. Thus there should be an equivalence:

"Snow is white" is true if and only if snow is white.

But there is nothing special about the sentence "snow is white" here. The same equivalence should be equally valid for any sentence. So the general form of the condition is

"A" is true if and only if A,

where the variable "A" can be replaced by any sentence.[7] This is Tarski's *T-scheme*. It is to be understood as a template which can be made into a truth definition for any sentence by replacing the symbol "A" with that sentence. Tarski proposes to use the T-scheme to characterize truth.

Actually, I ought to say that the sentence "snow is white" is very special in one important respect. Its meaning does not change at different times and places in the way that the meaning of a sentence like "it is raining here now" does. In this sense it is *universal*. This matters because, as I mentioned in the last section, when we are dealing with sentences that are not universal it generally does not make sense to talk about their truth without qualification. But since we are, in this book, principally interested in the universal statements of logic and mathematics, we can for the most part safely ignore complications that arise in the nonuniversal case.

Before we proceed, it is essential to understand that the "A" in the T-scheme is not the kind of variable one is used to working with. Usually a variable is thought of as ranging over a family of objects which it could be taken as referring to. For example, in the statement

If n is odd, then $n + 1$ is even,

the letter "n" might be a variable which ranges over the natural numbers. That is, it can be thought of as having a variable reference, i.e., as referring to any natural number, but importantly, it cannot be thought of as *being* any natural number. (Of course not; it is a letter.) Thus, a statement like "let n be a natural number" has the sense of "let the letter 'n' denote some natural number". This is seen in our ability to replace the variable "n" in the displayed sentence with any name for a natural number and get a meaningful statement. We could replace it with the numeral "37", say, and get

If 37 is odd, then $37 + 1$ is even.

The point is that what replaces the variable is a noun phrase, in this case a numeral, "37", which refers to the number 37.

In the T-scheme what replaces the variable is a sentence — not the name of a sentence, but the sentence itself. And as we discussed in Section 1.2, sentences do not have referents. Thus, since the variable "$\mathcal{A}$" in the T-scheme is to be replaced by something which has no referent, it is quite unlike the variable "n" in the preceding example. The T-scheme is not an assertion *about* arbitrary sentences in the way the preceding example is a statement about arbitrary natural numbers. It is more like a template which, for each sentence, can be used to generate an assertion involving that sentence. We can call "$\mathcal{A}$" a *schematic variable*, and write it in a calligraphic font, to emphasize this peculiarity. A variable that ranges over a family of objects in the usual way will be called *ordinary*. Where the ordinary variable "n" *represents* (say) a natural number, we can say that the schematic variable "$\mathcal{A}$" *serves as* (say) a sentence. I will call an expression that involves schematic variables a *schematic assertion* if it becomes a syntactically valid sentence when the schematic variables are replaced by the appropriate grammatical forms — in the present case, when "$\mathcal{A}$" is replaced by a sentence. This is why the T-scheme is a "scheme".

What does Tarski's T-scheme tell us? We may regard it as a clarification of the "correspondence" theory of truth, according to which a sentence is true if it corresponds to a state of affairs. Taken at face value, this theory raises as many questions as it answers, because we have to ask what is meant by a "state of affairs". Are these literally *things* that exist in the world? If so, what kind of things are they, and if not, then what is it that true sentences are supposed to correspond to? We might rather conclude that the correspondence theory involves figurative language whose real content is expressed by the T-scheme. Instead of saying that "snow is white" is true because *it corresponds to the fact that* snow is white, it is simpler and more cogent to say that "snow is white" is true because snow is white. Thus, "everything we need from, or even clearly understand in" the correspondence theory is captured by the T-scheme.[8]

Notwithstanding its success in this regard, the T-scheme's apparent lack of substance may at first seem a little disappointing. If there is nothing more to truth than this, then one could be forgiven for thinking that it must not be a very important notion. Truth as it is characterized by the T-scheme appears to be superfluous; seemingly, anything that could be said using truth could be said more directly without it.

However, this initial impression is very wrong. It is true that in ordinary discourse we sometimes ascribe truth merely for emphasis; this sentence is an example. This is not so interesting since such a use makes no substantive contribution to the content of the sentence. But that is not the only way truth is used. The reason it is so valuable is because it also enables us to make assertions indirectly.[9] This can be convenient: once we have given the Riemann hypothesis a name we can just say "suppose the Riemann hypothesis is true" without having to write the whole thing out. In this case, if necessary, the ascription of truth could be unpacked to a full statement of the assertion being referred to. But the real power of truth comes from cases where the assertions that let us make indirectly cannot be trivially unpacked. For instance, truth can be used to simultaneously affirm infinitely many sentences with a single sentence. We can say things like "every sentence that is formally derivable in S is true" where S is some axiomatic system. This kind of usage vastly increases our expressive power; it enables us to use a single sentence to effectively affirm an infinite conjunction of sentences — something of the form

$$\mathcal{A}_1 \text{ and } \mathcal{A}_2 \text{ and } \mathcal{A}_3 \text{ and } \cdots,$$

where in this case "$\mathcal{A}_1$", "$\mathcal{A}_2$", "$\mathcal{A}_3$", ... are to be replaced by the sentences that can be derived in S.

The liar sentence provides another, less happy, example of the kind of indirect assertion truth allows us to make. In this case, the sentence being referred to itself contains a truth ascription, in such a way that attempting to unpack it leads to an infinite regress.

We can also use truth to make global assertions, in the following manner. Consider the scheme "if $\mathcal{A}$, then $\mathcal{A}$", where "$\mathcal{A}$" is to be replaced by any sentence. All instances of this law can be simultaneously affirmed with the single sentence

Substituting any sentence for the schematic variable "$\mathcal{A}$"

in the scheme "if $\mathcal{A}$, then $\mathcal{A}$" produces a true sentence.

This statement can be formulated more efficiently using Quine's *quasi-quotation* notation.[10] In fact, schematic variables are not needed. Given any symbolic string A — that is, letting "A" be an ordinary variable which can denote any symbolic string — let $\ulcorner$if A, then $A\urcorner$ denote the string formed by concatenating the strings "if ", A, ", then ", and A, in that order. That is, it is read as a quotation but with each syntactic variable replaced by the string it represents. Thus for $A =$ "snow is white" we have $\ulcorner$if A, then $A\urcorner =$ "if snow is white, then snow is white". Then we can restate the preceding as

For any sentence A, the sentence $\ulcorner$if A, then $A\urcorner$ is true.

This single statement affirms all sentences which follow the pattern "if snow is white, then snow is white", with any sentence in place of "snow is white". That is what we can do using truth.

It may occur to the reader that infinite conjunctions can sometimes be asserted without invoking truth. For example, consider the sentence "for every natural number n, if n is odd then $n + 1$ is even". Is this not a single sentence that, without invoking truth, affirms an infinite conjunction, namely, the conjunction of the sentences "if n is odd then $n + 1$ is even" with "n" successively replaced by "0", "1", "2", ...? Yes, it certainly is. But that is only possible because here "n" is an ordinary variable, so that the quantifying expression "for every n" makes grammatical sense. The scheme "if $\mathcal{A}$, then $\mathcal{A}$" cannot be quantified in a similar way.[11] We cannot say "for every $\mathcal{A}$, if $\mathcal{A}$, then $\mathcal{A}$" because "$\mathcal{A}$" is merely a placeholder, a slot in a template, so that the phrase "for every $\mathcal{A}$" is nonsensical. "For every" demands to be followed by an ordinary variable: "for every natural number n" makes sense because n, the thing the letter "n" refers to, can *be* any number. Thus the only way to quantify over all sentences is by saying "for every sentence A" with "A" an ordinary variable. But "for every sentence A, if A, then A" is nonsensical because "if ... then" needs to combine with a pair of sentences, not a pair of noun phrases; and similarly "for every sentence A, $\ulcorner$if A, then $A\urcorner$" is nonsensical because $\ulcorner$if A, then $A\urcorner$ is not a sentence, but rather a noun phrase that refers to a sentence, and one cannot assert a noun phrase. The upshot is that we can generalize over ordinary variables without referencing truth, but we need truth to generalize over schematic variables.[12]

Let me say a little more about how it is that truth gives us this ability.

Another way to express the fact that truth enables us to assert sentences indirectly is to say that it is a device of *disquotation*.[13] We turn a sentence into a name for that sentence by enclosing it in quotation marks. This produces a syntactic object that can be manipulated but not asserted (because it is now a noun phrase). Calling the quoted sentence true, however, in effect removes the quotation marks and has the same force that asserting the original sentence would have had. The terms *use* and *mention* are helpful here. In " 'snow is white' is true if and only if snow is white", the first "snow is white" is mentioned and the second one is used. To mention a sentence is to refer to it as a symbolic string; to use it is to include it as a functioning component of an assertion. The point is that enclosing in quotation marks converts use into mention, and ascribing truth converts mention into use.

It should be easy to see that we can use the mention-to-use conversion aspect of truth to generalize over schematic variables of any kind, as claimed above. We simply have to describe an algorithm which generates all instances of a given template with appropriate grammatical structures inserted at the relevant places, and then say that every sentence produced by this algorithm is true.

So truth, as it is portrayed by the T-scheme, is not as empty a notion as it might first appear to be. On the contrary, it is a powerful tool for saying things that could not otherwise be said. But it is time for us to acknowledge a severe problem: if we need to use truth to affirm schematic assertions, how are we to affirm the T-scheme? It would seem that the T-scheme is useless as a definition of truth, because if we do not already possess a global notion of truth, we have no way to say that every instance of the T-scheme is true. I call this *the Tarskian catastrophe*.[14]

There are two fairly obvious possible responses to this difficulty. One is to abandon the idea of globally affirming the T-scheme, and instead take all the separate instances of the T-scheme as collectively defining truth. The other is to assume that we come to the table with a concept of truth already in hand, which we can then use to affirm the statement that all instances of the T-scheme are true. On the latter view we are not using the T-scheme to define a novel concept, but merely to describe a concept we already possess. Neither of these responses works; I will discuss them further in Sections 2.4 and 2.5. Tarski's own method of dealing with the problem will be considered in Section 3.7.

What we must not do is to mask the problem by surreptitiously employing a synonym for the word "true".[15] Saying that every instance of the T-scheme *holds*, or *is the case*, or *is satisfied* only makes sense if we already have such concepts, which would be tantamount to assuming that we already have a concept of truth. If that is the assumption then we had better own up to it (which is a respectable position, although, as I just mentioned, I do not think it succeeds).

The most pessimistic response to the Tarskian catastrophe is to simply abandon the whole idea of truth. In light of the liar paradox, this might not seem so unreasonable. However, before we adopt this alternative we ought to investigate just how badly the T-scheme is implicated in the liar paradox. In order to do this, let us first derive the contradiction a little more formally, in a way that brings out the role of the T-scheme. Thus, let L be the sentence "L is not true". (If that already seems questionable, recall that Quine's version of the paradox allows us to avoid the use of an explicit label, at the cost of minor complication.) Proving the implication "if L is true then L is not true" now takes two steps:

> If L is true then "L is not true" is true (definition of L);
> if "L is not true" is true then L is not true (T-scheme).

The argument for "if L is not true then L is true" is just the reverse deduction:

> If L is not true then "L is not true" is true (T-scheme);
> if "L is not true" is true then L is true (definition of L).

Plainly, it is the T-scheme — the equivalence between the assertion that a sentence is true and the sentence itself — that drives the paradox. No other aspect of truth, if there even are any, comes into play. But notably, one does not need the entire T-scheme, just the single instance where "$\mathcal{A}$" is replaced by "L is not true". This suggests that the problem is limited to certain very special, one might even say pathological, instances of the T-scheme.

Indeed, Tarski made a powerful case that the T-scheme is generally unproblematic when it is applied to an object language from some external perspective. He did this by showing how, in broad circumstances, one can explicitly define a *truth predicate* for a given language — intuitively, a predicate which, if it is substituted for the phrase "is true" in the T-scheme, makes every instance of that scheme true when "$\mathcal{A}$" replaced by any sentence of the object language. I say "intuitively" because this characterization is framed in terms of the *truth* of the T-scheme, which is obviously worrying. Despite the apparent clarity of the concept, it is not obvious how to say what constitutes a truth predicate for a given language without employing a truth predicate that applies even more broadly. This is just another manifestation of the Tarskian catastrophe. There is a real puzzle here, because, as we will see, we can make a compelling informal argument that Tarski's construction does show that the T-scheme can be made to work for a variety of languages.

I will describe three constructions of such localized notions of truth in the next three sections and present a general method for constructing truth predicates in Section 2.3. A limitation of this approach is that such a predicate only applies to sentences of the object language; in particular, it cannot be applied to sentences which contain the new predicate itself. Of course, the liar paradox would seem to show that some such restriction is inevitable.

1.4 Numerical truth

Numerical sentences. Informal meanings of $\wedge$ and $\vee$. Entailment as a relation between sentences. Reducing negation to implication. Definition of the truth function τ. The T-scheme for τ.

In this section I will show how to define a truth predicate for a very elementary type of sentence whose only content involves concrete numerical calculations with whole numbers. Then in the next section we will pass to the full language of number theory.

The relevant sentences contain no variables, and there is only a single constant symbol, "0". We also have three operation symbols, "$'$" (successor), "$+$" (sum), and "$\cdot$" (product). The idea is that "$+$" and "$\cdot$" are to be thought of as the usual sum and product operations, and "$'$" indicates the operation of adding one. Thus "0" represents the number 0, "$(0')$" the number 1, "$((0')')$" the number 2, and so on. I will use the notation "$\hat{n}$" to stand for the string formed by applying n sucessor symbols to "0" and parenthesizing appropriately. Thus $\hat{0} = $ "0", $\hat{1} = $ "$(0')$", $\hat{2} = $ "$((0')')$", etc., and $\hat{n}$ can be any of these. (So "$\hat{2}$" is a syntactic constant and "$\hat{n}$" is a syntactic variable.) Expressions of this form are *numerals*.

General numerical expressions, or *numerical terms*, can be built up using the three operation symbols in various combinations. The formal definition states that "0" is a numerical term, if t is a numerical term then so is $\ulcorner (t') \urcorner$, and if t_1 and t_2 are numerical terms then so are $\ulcorner(t_1 + t_2)\urcorner$ and $\ulcorner(t_1 \cdot t_2)\urcorner$. (Recall the quasi-quotation notation introduced on page 12.) This exhausts the possible ways to construct numerical terms.

For the sake of readability, I will omit unnecessary parentheses when writing terms by invoking the standard order of precedence: first associate all occurrences of "$'$", then all occurrences of "$\cdot$", reading left to right, and last all occurrences of "$+$", reading left to right. For example,

$$0 + 0' \cdot 0'' \qquad \text{and} \qquad 0 \cdot 0 \cdot 0$$

respectively abbreviate

$$(0 + ((0') \cdot ((0')'))) \qquad \text{and} \qquad ((0 \cdot 0) \cdot 0).$$

An *atomic numerical sentence* is any expression of the form $\ulcorner t_1 = t_2 \urcorner$ where t_1 and t_2 are numerical terms. It is to be understood as asserting that t_1 and t_2 represent the same number. As it was for terms, the characterization of general *numerical sentences* is recursive: whenever A is a numerical sentence, so is $\ulcorner(\neg A)\urcorner$, and whenever A and B are numerical sentences so are $\ulcorner(A \wedge B)\urcorner$, $\ulcorner(A \vee B)\urcorner$, and $\ulcorner(A \rightarrow B)\urcorner$. This is the complete definition: a numerical sentence is a syntactic expression that can be built up from the atomic numerical sentences using the *logical connectives* "$\neg$" (negation), "$\wedge$" (conjunction), "$\vee$" (disjunction), and "$\rightarrow$" (implication).

I will also use order of precedence rules to omit parentheses when writing sentences. The standard rules instruct us to first associate all occurences of "$\neg$", then all occurrences of "$\wedge$" and "$\vee$", reading left to right, and last all occurrences of "$\rightarrow$", reading left to right. For example,

$$\ulcorner A \wedge \neg B \rightarrow A \vee B \wedge C \urcorner \qquad \text{and} \qquad \ulcorner A \rightarrow B \rightarrow C \urcorner$$

respectively abbreviate

$$\ulcorner((A \wedge (\neg B)) \rightarrow ((A \vee B) \wedge C))\urcorner \qquad \text{and} \qquad \ulcorner((A \rightarrow B) \rightarrow C)\urcorner.$$

I may also occasionally add unnecessary parentheses, when doing so will improve readability.

Numerical sentences convey information about the results of performing simple numerical calculations. In order to explain what they mean, I have to say how to interpret the logical connectives.

The most straightforward connectives are "$\wedge$", conjunction, and "$\vee$", disjunction. The first is read as "and"; thus the ordinary language version of $\ulcorner A \wedge B \urcorner$ is $\ulcorner A$ and $B \urcorner$. To give a concrete example, $\ulcorner(\hat{1} + \hat{2} = \hat{3}) \wedge (\hat{2} + \hat{3} = \hat{5})\urcorner$ means "$1 + 2 = 3$ and $2 + 3 = 5$".

The connective "∨" is read as "or", so that $\ulcorner(\hat{1}+\hat{2}=\hat{3}) \vee (\hat{2}+\hat{3}=\hat{5})\urcorner$ means "$1+2=3$ or $2+3=5$". The one comment I need to add is that "or" is meant in the inclusive sense: $1+2=3$, or $2+3=5$, or both.

The sentence $\ulcorner A \to B\urcorner$ can be read as $\ulcorner$if A, then $B\urcorner$ or $\ulcorner$whenever A, also $B\urcorner$. The second formulation seems a little clearer because it avoids the suggestion that A is hypothetical. For example, $\ulcorner(\hat{1}+\hat{2}=\hat{3}) \to (\hat{2}+\hat{3}=\hat{5})\urcorner$ means "in every circumstance where $1+2=3$, also $2+3=5$". Now since everything we can express using numerical sentences is universal — $1+2$ does, in fact, always equal 3 — we are in a sort of a degenerate situation as regards the qualifier "whenever". But the interpretation is still cogent: either $1+2$ always equals 3 and $2+3$ always equals 5, or $1+2$ never equals 3 and it does not matter whether $2+3$ ever equals 5.

The ordinary language interpretation of implication deserves a brief digression. Many authors have noted that $\ulcorner A \to B\urcorner$ is sometimes ungrammatically read as $\ulcorner A$ implies $B\urcorner$. We can say "if $1+2=3$, then $2+3=5$" and we can also say "whenever $1+2=3$, also $2+3=5$", but "$1+2=3$ implies $2+3=5$" is ungrammatical because "implies" is an ordinary verb that needs to combine with two noun phrases, not two sentences. The correct formulation is either "that $1+2=3$ implies that $2+3=5$" or "the sentence '$1+2=3$' implies the sentence '$2+3=5$'". (So $\ulcorner$that A implies that $B\urcorner$ is synonymous with $\ulcorner$if A, then $B\urcorner$.)

This can be confusing because the sentence "A implies B" is grammatically correct when "A" and "B" are ordinary variables which function as noun phrases. (Notice the shift from quasi-quotation to quotation.) In other words, entailment acts as a relation between pairs of sentences. This relation is defined by the following condition:

> For all sentences A and B, A *implies* (or *entails*) B if and only if the sentence $\ulcorner$if A, then $B\urcorner$ is true.

It is important to note that this definition requires a truth predicate. Moreover, it is also possible to go in the reverse direction and use the entailment relation to define truth. Namely, we can say that a sentence is true if and only if it is implied by $\ulcorner\hat{0}=\hat{0}\urcorner$. Intuitively, this definition of truth works since $\ulcorner\hat{0}=\hat{0}\urcorner$ is true, so that the only way it can imply A is for A to be true. A more general way to write this is

> A is true if and only if "⊤" implies A,

where we introduce the symbol "⊤" into the language as an abbreviation of the sentence $\ulcorner \hat{0} = \hat{0} \urcorner$.

What this shows is that having entailment as a relation between sentences is equivalent to having a truth predicate. In Section 1.3 I warned against letting truth synonyms like "holds" or "is the case" sneak into a definition of truth, and now we need to add "implies" and "entails" to that list as well.[16] (But if we know what $\ulcorner$if A, then $B\urcorner$ means, surely we also know what $\ulcorner$"A" implies "B"$\urcorner$ means. Is having a truth predicate automatic, then, once we understand a language? See Section 1.7.)

Returning to our infomal explanation of the logical connectives, we still have to consider negation. Negation is expressed by the connective "$\neg$". This looks simple enough, but it is not as trivial as it seems. We cannot say that $\ulcorner \neg A \urcorner$ means $\ulcorner$not $A \urcorner$ since this is not grammatical. In ordinary language you cannot say "not $1 + 2$ equals 3". You can say "$1 + 2$ does not equal 3", which might suggest that "not" simply attaches to the verb, but more complex sentences refute this idea: the negation of "$1 + 2 = 3$ and $1 + 2 = 4$" is not "$1 + 2$ does not equal 3 and $1 + 2$ does not equal 4". Both statements are false. We will see later (Section 4.3) that there is a real issue here, an asymmetry between a sentence and its negation, that the standard gloss on "$\neg$" obscures.

The phrase "it is not the case that" is a more grammatical rendering of "$\neg$". Thus $\ulcorner \neg (\hat{1} + \hat{2} = \hat{3}) \urcorner$ can be read as "it is not the case that $1 + 2 = 3$", and this also works with more complex sentences. But this interpretation of "$\neg$" invokes the notion of "being the case", which is synonymous with being true. We could just as well read $\ulcorner \neg (\hat{1} + \hat{2} = \hat{3}) \urcorner$ as "it is not true that $1 + 2 = 3$". Since our central goal here is to understand truth, we must strive to avoid taking it as given, even at an informal level.

We can do this by reducing negation to implication. Namely, introduce the symbol "$\perp$" into the language as an abbreviation of the sentence $\ulcorner \hat{0} = \hat{1} \urcorner$. Then we can interpret $\ulcorner \neg A \urcorner$ as an abbreviation of $\ulcorner A \to \perp \urcorner$. The idea is that $\ulcorner$whenever A, also $\perp \urcorner$ effectively denies A.

This completes our informal explanation of the logical constants. Notice that I did not employ the notion of truth anywhere in this explanation. I did not say that $\ulcorner A \wedge B \urcorner$ is true if and only if A is true and B is true, only that $\ulcorner A \wedge B \urcorner$ means $\ulcorner A$ and $B \urcorner$. It is worth noting this because we are trying to get at the idea of truth from first principles, so we do not want to be referencing it in some unnoticed way. Of course, one already has to know what "and" means in order for our definition to make sense, so there is still a question of whether we might have needed to use a notion of truth when we first learned the informal metalanguage in which this whole discussion is couched. I will say more about the relation between truth and meaning in Sections 1.7 and 4.2.

Let us now construct a truth predicate for the family of numerical sentences. We will do this by defining a *truth function* τ which assigns the *truth value* 0 or 1 to each such sentence.

To start, given any atomic numerical sentence $\ulcorner t_1 = t_2 \urcorner$, the terms t_1 and t_2 can be numerically evaluated by simple algorithms that we learn in elementary school. Define $\tau(\ulcorner t_1 = t_2 \urcorner)$ to be either 1 or 0 according to whether t_1 and t_2 evaluate to the same or different numbers. Then recursively define τ for more complex sentences by the following rules:

$$\tau(\ulcorner \neg A \urcorner) = \begin{cases} 0 & \text{if } \tau(A) = 1 \\ 1 & \text{if } \tau(A) = 0 \end{cases}$$

$$\tau(\ulcorner A \wedge B \urcorner) = \begin{cases} 0 & \text{if } \tau(A) = 0 \text{ or } \tau(B) = 0 \\ 1 & \text{if } \tau(A) = \tau(B) = 1 \end{cases}$$

$$\tau(\ulcorner A \vee B \urcorner) = \begin{cases} 0 & \text{if } \tau(A) = \tau(B) = 0 \\ 1 & \text{if } \tau(A) = 1 \text{ or } \tau(B) = 1 \end{cases}$$

$$\tau(\ulcorner A \rightarrow B \urcorner) = \begin{cases} 0 & \text{if } \tau(A) = 1 \text{ and } \tau(B) = 0 \\ 1 & \text{if } \tau(A) = 0 \text{ or } \tau(B) = 1. \end{cases}$$

Evidently, the evaluation of $\tau(A)$, for arbitrarily complex A, is achieved by a simple computation.

Intuitively, $\tau(A) = 1$ says that A is true and $\tau(A) = 0$ says that A is false. For instance, intuitively, $\ulcorner A \wedge B \urcorner$ is true if and only if both A and B are true, and correspondingly $\tau(\ulcorner A \wedge B \urcorner) = 1$ if and only if both $\tau(A) = 1$ and $\tau(B) = 1$. Thus one might expect "is evaluated to 1 by τ" to function as a truth predicate. In fact it does, as we will now verify.

The claim is that for any numerical sentence A we can prove the sentence $\ulcorner$ "A" is evaluated to 1 by τ if and only if $A \urcorner$. Here the left side of the equivalence says that τ takes the value 1 on a certain symbolic string, and the right side is to be understood informally as an assertion about natural numbers, according to the informal interpretations of the logical constants I outlined above.

To put it more compactly, for every A I will show how to prove

$$\ulcorner \tau(\text{``}A\text{''}) = 1 \qquad \leftrightarrow \qquad A \urcorner,$$

where the logical connective "↔" (equivalence), to be read as "if and only if", is defined by taking $\ulcorner A \leftrightarrow B \urcorner$ to abbreviate $\ulcorner (A \to B) \wedge (B \to A) \urcorner$.

We can see how to prove the desired equivalence by induction on the complexity of A. That is to say, we prove it first for atomic sentences, and then for progressively more complex sentences.

In the atomic case, where A is $\ulcorner t_1 = t_2 \urcorner$ for some numerical terms t_1 and t_2, the statement to be proven says that $\tau(A) = 1$ if and only if t_1 and t_2 represent the same number. For instance, $\tau(\ulcorner \hat{1} + \hat{2} = \hat{3} \urcorner) = 1$ if and only if $1 + 2$ equals 3. This is an immediate consequence of the way we defined τ on atomic sentences: by definition $\tau(\ulcorner t_1 = t_2 \urcorner) = 1$ if and only if the terms t_1 and t_2 evaluate to the same number.

In the case of a conjunction, we have from the definition of τ that $\tau(\ulcorner A \wedge B \urcorner) = 1$ if and only if $\tau(A) = \tau(B) = 1$. So if we have already proven

$$\ulcorner \tau(``A") = 1 \qquad \leftrightarrow \qquad A \urcorner \qquad\qquad (*)$$

and

$$\ulcorner \tau(``B") = 1 \qquad \leftrightarrow \qquad B \urcorner \qquad\qquad (**)$$

for some particular A and B, we can deduce

$$\ulcorner \tau(``A \wedge B") = 1 \qquad \leftrightarrow \qquad A \text{ and } B \urcorner$$

for that A and B. This shows how to prove the desired equivalence for a conjunction once we have proven it for each of the two constituent sentences.

Similarly, for all A and B we can prove that

$$\tau(\ulcorner A \vee B \urcorner) = 1 \qquad \text{if and only if} \qquad \tau(A) = 1 \text{ or } \tau(B) = 1$$

and

$$\tau(\ulcorner A \to B \urcorner) = 1 \qquad \text{if and only if} \qquad \text{whenever } \tau(A) = 1, \text{ also } \tau(B) = 1;$$

combining these with $(*)$ and $(**)$ yields

$$\ulcorner \tau(``A \vee B") = 1 \qquad \leftrightarrow \qquad A \text{ or } B \urcorner$$

and

$$\ulcorner \tau(``A \to B") = 1 \qquad \leftrightarrow \qquad \text{whenever } A, \text{ also } B \urcorner.$$

So we can also prove the desired equivalence for a disjunction and for an implication, once we have proven it for each of the two constituent sentences.

Negation does not have to be fully analyzed separately since we have informally interpreted $\ulcorner \neg A \urcorner$ as $\ulcorner A \to \bot \urcorner$. We just have to observe that for all A we can prove

$$\ulcorner \tau(\text{``}A \to \bot\text{''}) = 1 \qquad \leftrightarrow \qquad \tau(\text{``}A\text{''}) = 0 \urcorner$$

and that combining this with $(*)$ yields

$$\ulcorner \tau(\text{``}A \to \bot\text{''}) = 1 \qquad \leftrightarrow \qquad \text{whenever } A, \text{ also } 0 = 1 \urcorner.$$

This shows how to prove the desired equivalence for a negation once we have proven it for the constituent sentence, and that completes the argument. We have seen how to verify the claimed equivalence for numerical sentences of arbitrary complexity.

However, when we try to encapsulate what it is we have proven, or at least shown how to prove, in a single statement, the Tarskian catastrophe intervenes. We would like to say that every sentence of the form $\ulcorner \tau(\text{``}A\text{''}) = 1 \leftrightarrow A \urcorner$ is *true*, with $A$ ranging over all numerical sentences, but saying this would require a truth predicate that applies both to all statements of the form $\ulcorner \tau(\text{``}A\text{''}) = 1 \urcorner$ and to all numerical sentences. No doubt we have such a predicate. However, the whole point of the present discussion is to show how we can construct a truth predicate for the numerical sentences, so invoking a prior possession of truth predicate with even broader scope would defeat the purpose.

Nonetheless, it seems clear that τ really is a truth predicate for the numerical sentences, if only there were some way to say this without invoking truth. We know that we can prove every instance of the T-scheme, and moreover, each such instance can be directly asserted without invoking truth. The need for truth only comes in when we try to assert all of these instances at once.

Maybe the problem could be solved if we were allowed to use infinitely long sentences. Then we could simultaneously affirm every instance of the T-scheme by enumerating the numerical sentences as A_1, A_2, A_3, ... and asserting the single infinite sentence

$$\ulcorner (\tau(\text{``}A_1\text{''}) = 1 \leftrightarrow A_1) \wedge (\tau(\text{``}A_2\text{''}) = 1 \leftrightarrow A_2) \wedge \cdots . \urcorner \tag{$\dagger$}$$

This formulation makes no overt reference to truth. We would have to treat the argument we made above as an algorithm for generating a sequence of proofs, one for each conjunct of ($\dagger$). While this approach may seem somewhat exorbitant, at least it shows that we can formulate the fact that τ is a truth predicate in a way that has real, noncircular content.

Or does it? Since we cannot literally write down the infinite conjunction in (†), we seem not to have any way to directly assert it. Recall that one of the functions of a truth predicate is to enable us to affirm infinite conjunctions. So it seems that we need truth in order to assert (†), and thus the Tarskian catastrophe is not so easily defeated. But see Section 2.5 for more on this point.

One other comment. Since, as we have noted, the value of τ on any numerical sentence can be straightforwardly computed, it is easy to convince oneself that there is a computer program P such that a numerical sentence A will be P-accepted if and only if $\tau(A) = 1$. This bears on the question of whether truth can be a property of syntactic objects like sentences, but I will postpone that discussion to the end of Section 1.6.

1.5 Arithmetic

*The language of arithmetic. Quantifiers. Definition of the truth
function τ. The T-scheme for τ. Infinite computations.*

Let us consider a more sophisticated language, the language of arithmetic. Here "arithmetic" is meant not in the elementary school sense of numerical calculation, but rather in the professional mathematician's sense of "the branch of mathematics that deals with the natural numbers".

This language is built up from an infinite number of variables "x", "y", "z", ... and a single constant symbol "0". The variables are to be thought of as ranging over the natural numbers $\mathbb{N} = \{0, 1, 2, \ldots\}$. A *term* is any expression built up from 0 and the variables using the operation symbols "$'$", "$+$", and "$\cdot$", in the same way as for numerical terms, but now also allowing variables to appear. An *atomic formula* is any expression of the form $\ulcorner t_1 = t_2 \urcorner$ where t_1 and t_2 are terms; note that if any variables appear in t_1 or t_2, then the sense of the formula can vary depending on the values assigned to those variables. All *formulas* of arithmetic are then built up from the atomic formulas following the same rules as for numerical sentences, using the logical connectives "$\neg$", "$\wedge$", "$\vee$", and "$\rightarrow$", together with two new clauses corresponding to the quantifiers "$\forall$" and "$\exists$". The additional clauses state that if A is a formula then $\ulcorner (\forall x)A \urcorner$ and $\ulcorner (\exists x)A \urcorner$ are also formulas, and similarly for other variables besides x.

(The proper way to express that would be to use a metavariable "v" which ranges over the variables of arithmetic and to say that $\ulcorner (\forall v)A \urcorner$ and $\ulcorner (\exists v)A \urcorner$ are formulas for any formula A and any variable v. But for the sake of readability I will avoid using metavariables and expect the reader to understand that variables can be renamed at our discretion.)

A new feature here is that not every formula is considered a sentence. Say that an appearance of the variable "x" is *bound* if it lies within the scope of some "$(\forall x)$" or "$(\exists x)$" quantifier and *free* if it does not. For example, in the formula "$(\exists y)(x = y)$" the "x" is free and the "y" is bound. A *sentence* is a formula with no free variables.

Any formula can be turned into a sentence by replacing each free variable with a numeral — possibly different numerals for different variables, but the same numeral for different occurrences of the same variable. Or, more generally, we could replace each free variable with a numerical term. I will write "$A[t]$" for the formula obtained by replacing all free occurrences of the variable "x" with the term t, where the variable to be replaced would have been identified by previously writing "$A[x]$". (The notation does not imply that x actually does appear freely, or at all, in A. Nor do we require that t be a numerical term.)

The quantifiers "$\forall$" and "$\exists$" are read as "for all" and "for some" (or "there exists"). Thus $\ulcorner(\forall x)A\urcorner$ means $\ulcorner$for all x, $A\urcorner$ and $\ulcorner(\exists x)A\urcorner$ means $\ulcorner$for some x, $A\urcorner$. The logical connectives are interpreted in the same way as in the numerical case.

Formulas containing free variables can also be assigned a definite sense by using the convention that all free variables are to be understood as implicitly being universally quantified. According to this convention, the formula "$x + y = y + x$", standing alone, is understood to say that $x + y = y + x$ for all x and y. This convention fits well with the interpretation of "$\rightarrow$": for instance, the formula "$x = y \rightarrow y = x$" means "whenever $x = y$, also $y = x$", where "whenever" now encompasses the variety of ways of assigning values to "x" and "y".

The language of arithmetic is quite elementary, but we can express surprisingly sophisticated ideas with it. We can say that x is prime by saying that x is not equal to 1 but whenever $x = y \cdot z$, either $y = 1$ or $z = 1$. We can express the conjecture that every even number greater than 2 is a sum of two primes: x is even if there exists y such that $x = 2 \cdot y$, x is greater than 2 if there exists z such that $x = z + 3$, and so on.

Let us define a truth function τ for the language of arithmetic. It will assign the truth value 0 or 1 to each sentence of the language. As in the numerical case, the definition is recursive. We start with the atomic sentences, i.e., the atomic formulas which contain no variables. These are precisely the atomic numerical sentences, and we define τ on them in the same way we did this in Section 1.4. For sentences of greater complexity, we handle the logical connectives in the same way that we handled them in Section 1.4. Thus $\tau(\ulcorner \neg A \urcorner)$ is 0 if $\tau(A) = 1$, or 1 if $\tau(A) = 0$, and so on. Quantifiers are dealt with as follows. If $A[x]$ is a formula with no free variables other than, say, "x" — this is the only way that $\ulcorner (\forall x)A[x] \urcorner$ could be a sentence — then for each numeral $\hat{n}$ the formula $\ulcorner A[\hat{n}] \urcorner$ is a sentence. Thus when we go to define $\tau(\ulcorner (\forall x)A[x] \urcorner)$ and $\tau(\ulcorner (\exists x)A[x] \urcorner)$ we can assume that we have already defined $\tau(\ulcorner A[\hat{n}] \urcorner)$ for all n. Then we set $\tau(\ulcorner (\forall x)A[x] \urcorner) = 1$ if $\tau(\ulcorner A[\hat{n}] \urcorner) = 1$ for every n, or 0 if there is some n for which $\tau(\ulcorner A[\hat{n}] \urcorner) = 0$, and likewise, we set $\tau(\ulcorner (\exists x)A[x] \urcorner) = 1$ if there is some n for which $\tau(\ulcorner A[\hat{n}] \urcorner) = 1$, or 0 if $\tau(\ulcorner A[\hat{n}] \urcorner) = 0$ for every n.

This completes the definition of τ. Notice that when restricted to numerical sentences it agrees with the definition given in Section 1.4.

I claim that "is evaluated to 1 by τ" functions as a truth predicate for the language of arithmetic. For any arithmetical sentence A, we can prove the sentence $\ulcorner$ "A" is evaluated to 1 by τ if and only if $A \urcorner$, or more compactly,

$$\ulcorner \tau(``A") = 1 \qquad \leftrightarrow \qquad A \urcorner.$$

As before, the left side says that τ takes the value 1 on a certain symbolic string, and the right side is an informally understood assertion about natural numbers.

The argument follows the same lines as the analogous argument given in Section 1.4. When $A = \ulcorner t_1 = t_2 \urcorner$ is an atomic sentence, the proof that $\tau(\ulcorner t_1 = t_2 \urcorner) = 1$ if and only if $t_1 = t_2$ is immediate from the definition of τ, just as before. The logical connectives are also dealt with exactly as before. No more needs to be said about them here since the previous argument carries over verbatim. Finally, for quantified sentences, we have from the definition of τ that for any formula $A[x]$ with no free variables besides "x" we can prove that

$$\tau(\ulcorner (\forall x)A[x] \urcorner) = 1 \qquad \text{if and only if} \qquad \tau(\ulcorner A[\hat{n}] \urcorner) = 1 \text{ for all } n$$

and

$$\tau(\ulcorner (\exists x)A[x] \urcorner) = 1 \qquad \text{if and only if} \qquad \tau(\ulcorner A[\hat{n}] \urcorner) = 1 \text{ for some } n.$$

So if we have already proven for such a formula A each of the sentences

$$\ulcorner \tau(``A[\hat{n}]") = 1 \qquad \leftrightarrow \qquad A[\hat{n}] \urcorner$$

for every n, then we can deduce the sentences

$$\ulcorner \tau(``(\forall x)A[x]") = 1 \qquad \leftrightarrow \qquad A[x] \text{ for all } x \urcorner$$

and
$$\ulcorner \tau(``(\exists x)A[x]") = 1 \qquad \leftrightarrow \qquad A[x] \text{ for some } x \urcorner.$$
This shows how to prove the desired equivalence for a quantified sentence once we have proven it for each instance of the sentence, and that completes the argument.

A new issue that arises in the arithmetical case is the apparent need for infinitely long procedures. Already in the definition of τ, when it comes to evaluating τ on a quantified sentence we have to inspect its values on an infinite family of simpler sentences ("simpler" in the sense of having one fewer quantifier). In order to handle multiple quantifiers we have to be prepared to allow nested infinite computations. Also, in our verification of the T-scheme, when quantifiers appear we assume that we already have a proof of the scheme for the sentence $\ulcorner A[\hat{n}] \urcorner$, for every n. So we are dealing with infinite proofs, too.

I anticipate that most readers, but not all, will be willing to accept the theoretical possibility of such computations and proofs. This is one reason for considering the numerical case first; in that setting everything was finite. We will return to the issue of infinitary reasoning later, but for now let us simply proceed with the understanding that a good deal of what we have to say about arithmetic on its face relies on the assumption that such reasoning is valid.

If this assumption is granted, then a version of the comment that I made at the end of Section 1.4, about there being a program P such that a sentence A will be P-accepted if and only if $\tau(A) = 1$, can be made here, as well. We have to generalize our concept of "computer program" to allow infinite computations, however.

Just as in Section 1.4, we have a problem formulating a single sentence that expresses the fact that τ is a truth predicate for arithmetic. It seems that this can only be done with the aid of a concept of truth that is broader than the one we are attempting to define. But, again, the claim that τ is a truth predicate for arithmetic is clearly right in some sense. The argument we made above intuitively shows that τ assigns the truth value 1 to those, and only those, sentences which are true statements about the natural numbers. This is a real, substantive fact about τ. For example, if $A[x]$ is "$(\exists y)(x = \hat{2} \cdot y)$" then $\tau(A[\hat{n}])$ is 1 when n is even and 0 when n is odd. It is especially galling that we have no trouble formulating this particular instance of the fact that τ is a truth predicate, or any other particular instance. The problem only comes when we try to assert all instances simultaneously.

1.6   Arithmetic mod n

Interpreting arithmetic mod n. Definition of the truth function
$\tilde{\tau}$. The T-scheme for $\tilde{\tau}$. Truth as a property of sentences.

Although our main goal is to understand truth in the most general
possible setting, in order to fix ideas it can be helpful to be able to consult
a specific formal language. The language of arithmetic is not necessarily
the best choice for this purpose, because the issue of infinite computations
mentioned above is liable to become a distraction. For this reason, it will
be good to have a version of arithmetic in which we do not have to worry
about infinite computations. *Arithmetic mod n* provides this for us.

Fix a natural number $n \geq 2$. The language of arithmetic mod n is
identical to the language of ordinary arithmetic, with the exception that
we replace the symbol "$=$" with "$\equiv$". Otherwise, we have precisely the same
repertoire of basic symbols, the terms are the same as before, and formulas
are built up from terms in the same way as before. The real difference
lies in the way the terms are interpreted. We now regard the variables as
ranging over the numbers from 0 to $n - 1$, inclusive, and the arithmetical
operations (successor, sum, and product) are understood as before except
that after any computation we retain only the remainder after dividing by
n.

For instance, when $n = 10$ this amounts to only retaining the ones digit
after any computation. Thus $\hat{1}\hat{3} \equiv \hat{3}$ and $\hat{9} \cdot \hat{4} \equiv \hat{3}\hat{6} \equiv \hat{6}$ in this case.

We can define a truth function $\tilde{\tau}$ for arithmetic mod n in exactly the
same way this was done for ordinary arithmetic, except that now it can be
evaluated by a finite computation for each formula, because when dealing
with quantifiers we do not have to check every natural number, only the
numbers 0 through $n - 1$. Note that $\tilde{\tau}$ can disagree with τ; for instance,
$\tau(\ulcorner \hat{1}\hat{3} = \hat{3} \urcorner) = 0$ but $\tilde{\tau}(\ulcorner \hat{1}\hat{3} \equiv \hat{3} \urcorner) = 1$ for $n = 10$.

Just as in the preceding two sections, for any sentence A in the language
of arithmetic mod n we can prove the T-scheme instance

$$\ulcorner \tilde{\tau}(``A") = 1 \qquad \leftrightarrow \qquad A \urcorner.$$

But as before, it is not obvious how to get across the idea that $\tilde{\tau}$ functions
as a truth predicate in a non-question-begging way.

I hope it is clear that in the setting of arithmetic mod n it is perfectly
cogent to consider *being evaluated to 1 by $\tilde{\tau}$* to be a property of sentences,
not propositions; and the same can be said for ordinary arithmetic, at least
if one accepts the idea of infinite computations. In the natural language

setting, perhaps it is not so hard to imagine an informal algorithm that operates on sentences of the form $\ulcorner$(concrete object) is (color word)$\urcorner$ and returns a truth value, and this might provide an explanation of what it could possibly mean for the symbolic string "snow is white" to be true. It might simply mean that this string elicits a certain response from a certain algorithmic procedure. (Of course, there could be a variety of ways to evaluate the truth of a sentence, but this is no more of a problem than the fact that a property could have a variety of equivalent definitions.)

At any rate, it should be evident that there is nothing wrong with dealing directly with the truth values $\tilde{\tau}$ takes on sentences. No detour through propositions is necessary.

1.7 Truth and meaning

Dummett's dilemma. The fallacy. Two equations in two unknowns.

Dummett raises the following problem. An instance $\ulcorner$"A" is true if and only if $A$$\urcorner$ of the T-scheme, for some particular sentence A, cannot tell us what $\ulcorner$"A" is true$\urcorner$ means if we do not first know what A means. But it seems that understanding what A means entails already knowing what conditions make it true; indeed, it is even plausible that understanding its meaning consists precisely in this. So it looks as though understanding the statement that tells us what it is for A to be true requires us to already know what it is for A to be true. We could not "first know what it was for a sentence to have the meaning that it does and then go on further to enquire what it is for it to be true".[17]

Clearly, the T-scheme cannot be used both to say what A means and to say what $\ulcorner$"A" is true$\urcorner$ means. This has been compared to trying to use a single equation to solve for two unknowns.[18] Still, one might have thought that an instance of the T-scheme could be used to define what it means for a particular sentence to be true, but if Dummett's criticism is correct, it cannot be employed for this purpose.

However, Dummett's criticism is not correct. The fallacy is simple but instructive. Saying that we gain an understanding of the meaning of "snow is white" from an explanation of the conditions under which "snow is white" is true is needlessly indirect. We should rather say that we gain an understanding of the meaning of "snow is white" from an explanation of the conditions under which snow is white. An example from mathematics: we gain an understanding of the meaning of "36 is even" from an explanation that 36 is even if it is divisible by 2, not from an explanation that "36 is even" is true if 36 is divisible by 2. If we have a truth predicate then the second formulation makes sense and is interchangeable with the first, but if we do not have a truth predicate then the first formulation does just fine by itself. The general principle that "we determine the sense of a sentence by laying down the conditions under which it is true"[19] may or may not be right — we will get to that question later — but the point is that this general principle makes an appeal to truth that is superfluous in individual cases. The same idea can be expressed schematically without using truth in the form

> We determine the sense of "$\mathcal{A}$" by laying down the conditions under which $\mathcal{A}$,

where "$\mathcal{A}$" can be replaced by any sentence. With the aid of truth this scheme can be expressed by the single statement that

> For each sentence A, we determine the sense of A by laying down the conditions under which A is true.

Remember that one of the main functions of truth is to allow us to convert schemes into sentences.

Since confusion about this issue seems to be nearly universal, I will risk laboring the point. We need to use truth to express the general principle that explains how we determine the sense of a sentence. At least, we need truth to accomplish this with a single sentence; we can also do it schematically without truth. But we emphatically do not need truth to explain how we determine the sense of any particular sentence. So we can *first* determine the sense of "snow is white" by laying down the conditions under which snow is white, and *then* determine the sense of " 'snow is white' is true" by saying " 'snow is white' is true if and only if snow is white". There is absolutely no reason we would need to already know what it is for "snow is white" to be true in order to understand the statement that tells us what it is for "snow is white" to be true.

The appearance of truth in Dummett's characterization of meaning is a red herring. Meaning is not about "truth" in some deep way. The point of using truth in this setting is to express a global regularity. It comes in when we wish to discuss the general process of endowing sentences with meaning, not when we perform the act of endowing any particular sentence with meaning.

Dummett's fallacy lies in assuming that if a general definition involves truth then every instance of that definition must involve truth. This need not be the case; it could be that truth was only used as a tool for expressing the general definition in the form of a single sentence.

For the sake of illustration, imagine a person who has no prior exposure to any kind of mathematics, but possesses enough English to be able to follow a step-by-step explanation of how to compute values of $\tilde{\tau}$ for, say, arithmetic mod 10. It may be helpful if we also assume that the word "true" does not already appear in this person's vocabulary, so that when they learn about truth for arithmetic mod 10 there will be no confusion with a previously understood truth predicate.

How would we teach such a person arithmetic mod 10? This might be done by first defining the syntactic structure of the language, then explaining how to evaluate $\tilde{\tau}$ on the sentences of the language, and finally introducing the scheme

$$\mathcal{A} \text{ if and only if } \tilde{\tau}(\text{``}\mathcal{A}\text{''}) = 1, \tag{1}$$

where the variable "$\mathcal{A}$" can be replaced by any sentence of arithmetic mod 10, in order to specify the meanings of those sentences. Here the right side is what was learned in the previous step, and we are now saying what the sentences mean by effectively reducing their content to the calculation of truth values. As promised, each instance of (1) establishes the meaning of a sentence directly, without referencing truth. Since (1) is a scheme, it must be understood as a device for generating, for each sentence, a characterization of the meaning of that sentence.

We could also, in parallel, define truth for arithmetic mod 10 by saying

$$\text{For all } A, A \text{ is true if and only if } \tilde{\tau}(A) = 1. \tag{2}$$

For instance, from this statement and an understanding of $\tilde{\tau}$ one has the ability to deduce laws like "the sentence $\ulcorner A \to A \urcorner$ is true for every sentence A of arithmetic mod 10". Alternatively, we could introduce truth via the T-scheme in conjunction with (1), but absent some device for globally affirming all instances of this scheme we could not go on to infer the global principle (2), only each of its instances. This is just the Tarskian catastrophe again.

Thus, the conditions which are used to assign meanings to sentences are also the conditions which are used to determine their truth,[20] but the two steps, which enable us to use the new sentences assertively and to discuss their truth, are independent of each other. In the metaphor mentioned earlier, we have solved Dummett's dilemma by using two equations, (1) and (2), to solve two unknowns.

The fact that (1) is schematic and (2) is not can be seen as an order-of-quantifiers issue. An understanding of a language may be shown in the ability to produce, for each sentence, a translation of that sentence into another, previously understood, language. An understanding of a truth predicate may be shown in the ability to globally state truth conditions for all sentences. This should help to illuminate how it is that truth predicates can carry substantive content. We might say that having a truth predicate for a language is about grasping an overall regularity that goes beyond separately understanding the meanings of the individual sentences.

The flip side of this observation is that having a truth predicate for only a finite set of sentences gives one no information at all beyond what is automatically available once one understands the language. Indeed, truth for a finite family of sentences could be characterized by a finite version of the sentence (†) described in Section 1.4, and this would amount to nothing more than an ordinary definition of a new concept symbol in terms of concepts that were already understood. It is precisely the infinitary aspect of (†) that gives it substantive content. This point may become clearer in Section 3.7.

This is also why the entailment relation carries more information than any of the sentences to which it applies (as it must, since it encodes the same information as a truth predicate, as we discussed in Section 1.4). This relation embodies global information about which statements of the form $\ulcorner$if A, then $B\urcorner$ are true, whereas to understand the language one merely has to understand $\ulcorner$if A, then $B\urcorner$ for each A and B separately.

Chapter 2

Concepts

2.1 Predicates and concepts

Concepts. Falling under. Predicates not referential. Expression. The case of arithmetic. Relations. Sets.

In Chapter 1 we considered the problem of defining truth predicates in some special settings related to arithmetic. We must now ask whether truth, in general and without qualification, is a well-defined concept. Before addressing this question, it may help to spend some time discussing what is meant by the word "concept".

In the philosophy of language, *concepts* (or *properties*) are alleged abstract correlates of predicates. As we discussed in Section 1.2, sentences allegedly express abstract entities called propositions. The relation between predicates and concepts is supposed to be analogous to the relation between sentences and propositions.[1] The idea is that a predicate is something that can be combined with a noun phrase, and a concept is something that in some analogous way can be combined with, or applied to, an object.

Concepts are fundamental to Frege's theory of language.[2] Frege is responsible for the modern understanding of mathematical language according to which complex sentences are built up from simpler sentences using logical connectives and quantifiers, in a manner exemplified by the language of arithmetic (Section 1.5).[3] In addition, according to Frege, the simplest sentences of a language, the atomic sentences, always relate objects to concepts. For instance, in the atomic sentence "snow is white", snow is the object and whiteness, or being white, is the concept which is ascribed to it.

Passing to the language of arithmetic, an atomic sentence like $\ulcorner \hat{2} = \hat{2} \urcorner$ can be analyzed into an object, the number 2, and a concept, equalling 2. Here I am using the word "object" in a generalized sense that includes abstract "objects" like numbers.

Whereas we say that the predicate "is white" *holds of*, or *is true of*, snow, we say that snow *falls under* the concept of whiteness (or that it *has* the property of whiteness). The natural identity criterion for concepts is: two concepts are the same if the same objects fall under them. Thus the concept of being a prime number less than 10 is the same as the concept of equalling 2, 3, 5, or 7. Another way to say this is that two predicates represent the same concept if they are true of the same objects. (The appearance of truth here is a possible red flag; we will return to this point in Section 2.4.)

Having a clear identity criterion for concepts is essential if we intend to reason about them rigorously. It also raises the possibility of a nominalistic interpretation of concept talk which regards concepts as literally being predicates but with a weakened notion of equality. That is, a nominalist might say that when we talk about concepts we are really just talking about predicates, but regarding two predicates as effectively "the same" if they hold of the same objects. Similarly, a nominalist can say that when we talk about propositions we are really just talking about sentences, but regarding two sentences as "the same" if each implies the other.

What we cannot do is to regard concepts as referents of predicates, for the same reason we cannot regard propositions as referents of sentences (as we discussed in Section 1.2).[4] Predicates are not noun phrases and substituting them for noun phrases yields nonsensical results. We can say " 'is white' is a predicate" because enclosing "is white" in quotation marks turns it into a name. But "is white is a predicate" does not express a coherent thought. Predicates do not refer to anything.[5]

By the same token, just as the *that X* construction nominalizes a sentence, we have nominalizing constructions for predicates such as *Xness* and *being X*. So, for instance, whiteness is the concept associated to the predicate "is white". As I have just emphasized, the predicate "is white" does not *name* the concept of whiteness (as the word "whiteness" does). It might rather be said to *express* that concept in much the same way that sentences express propositions. Thus, saying that a predicate expresses a concept C is equivalent to saying that the sentence formed by combining the name for any object x with that predicate expresses the proposition that x falls under C.

With minor modifications, all of these ideas also apply to formal languages. Consider the case of arithmetic (Section 1.5). Here the terms of the language play the role of noun phrases, and we may consider the "predicates" of the language to be the formulas that have one free variable. Two such formulas $A[x]$ and $B[y]$ express the same concept if for every n the equivalence $\ulcorner A[\hat{n}] \leftrightarrow B[\hat{n}]\urcorner$ is true, or, avoiding explicit mention of truth, if

$$\tau(\ulcorner A[\hat{n}]\urcorner) = \tau(\ulcorner B[\hat{n}]\urcorner)$$

for all n. For example, being even and being prime are concepts expressible in the language of arithmetic, via the respective formulas

$$(\exists z)(x = \hat{2} \cdot z)$$

and

$$\neg\, y = \hat{1} \quad \wedge \quad (\forall z)(\forall w)(y = z \cdot w \quad \rightarrow \quad z = \hat{1} \vee w = \hat{1}).$$

So is the concept of being a sum of four squares, which by a famous theorem of Lagrange is the same as the concept of being a natural number.

It is straightforward to generalize the above to formulas with more than one free variable. For instance, formulas with two free variables (taken in a specified order) are said to express *relations*. The identity criterion we use here says that two such formulas express the same relation if and only if they are true of the same pairs of objects. Formulas with more than two free variables may be regarded as expressing generalized relations, but there is less need to invent a special terminology for them.

So formulas with two free variables express relations, formulas with one free variable express concepts, and we already decided in Section 1.2 that formulas with no free variables, i.e., sentences, express propositions. Note that the identity criterion we are using now fits with the convention we adopted in Section 1.2 according to which two sentences express the same proposition if and only if each implies the other. For the natural analog of $\tau(\ulcorner A[\hat{m}, \hat{n}]\urcorner) = \tau(\ulcorner B[\hat{m}, \hat{n}]\urcorner)$ for all m and n (identity for relations) and $\tau(\ulcorner A[\hat{n}]\urcorner) = \tau(\ulcorner B[\hat{n}]\urcorner)$ for all n (identity for concepts) is simply $\tau(A) = \tau(B)$ (identity for propositions), and this is equivalent to saying that A and B imply each other.

Since every sentence of arithmetic is either true or false, the language of arithmetic accomodates exactly two propositions. This probably goes against whatever naive intuition we have for propositions, but for our purposes the definition is still good because we want the identity criterion for propositions to follow the same pattern as for concepts and relations.

Another natural idea is to associate concepts with sets. For instance, in the case of arithmetic, if $A[x]$ is any formula with one free variable x then the set of n for which $\ulcorner A[\hat{n}]\urcorner$ is true — for which $\tau(\ulcorner A[\hat{n}]\urcorner) = 1$ — is a set of numbers that characterizes the concept expressed by A, in the sense that another formula B with one free variable expresses the same concept if and only if it gives rise to the same set. That is, we know what concept is expressed by a predicate if we know the set of objects for which that predicate holds, and vice versa.

(The general pattern is that relations correspond to subsets of $\mathbb{N}^2$, concepts correspond to subsets of $\mathbb{N}^1 \cong \mathbb{N}$, and propositions correspond to subsets of the one-element set $\mathbb{N}^0 = \{\emptyset\}$, of which there are two — $\emptyset$ and $\{\emptyset\}$ — representing falsehood and truth.)

This association between concepts and sets is perfectly reasonable in some special settings, but it cannot be accepted as a general principle. In broader settings there are concepts that do not correspond to sets in the above manner. For example, if we are able to talk about sets in general then being a set must itself be a legitimate concept, but unless we radically alter our intuitive notion of a set, we know that there is no set of all sets that could correspond to this concept. So not every concept corresponds to a set.

Since the collection of all sets is not a set but a proper class, should we instead associate concepts with classes? No, this generates exactly the same difficulty for the concept of being a class. If being a class is a legitimate concept then it cannot itself correspond to any class because there is no class of all classes.

The nature of the relationship between sets, classes, and concepts is something I will attempt to clarify much later, in Chapter 6. By that point it should have become evident that concepts constitute a broader category than either sets or classes. In particular, we will see in Chapter 5 that in order for there to be a concept of being a concept, we have to allow for the possibility that concepts can fail to satisfy the law of excluded middle. That is, there will be a concept for which we are not in a position to affirm that every object definitely either does or does not fall under it. Thus, it would be a serious error for us to assume at this point that concepts correspond to sets in the manner suggested above.

2.2   Russell's paradox

Truth versus holding. Indirect ascription. The T-scheme for holding. Russell's paradox. Set and class versions of the paradox.

There is a close relationship between saying that a sentence (or proposition) is true, on one side, and saying that a predicate holds of an object (or an object falls under a concept), on the other. Let us explore this relationship.

To start with, the holding relation can be defined in terms of truth. We can say

> The predicate A holds of a if and only if any sentence formed by combining a name for a with A is true.

The formal language version of this condition is

> The formula $A[x]$ holds of a if and only if the sentence $A[\hat{a}]$ is true,

where A is a formula with no free variables besides x and $\hat{a}$ can be any name for a. This shows that we can use knowledge of which sentences are true to determine which predicates hold of which objects.

To make the corresponding definition at the more abstract level of concepts, we need a notation for the proposition which ascribes a concept C to an object x. I will denote this proposition by "$C[[x]]$". That is, $C[[x]]$ is the proposition expressed by any sentence formed by combining a name for x with a predicate that expresses C. Then falling under is characterized by the condition

> The object x falls under the concept C if and only if the proposition $C[[x]]$ is true.

Again, truth figures essentially in this statement. We can determine which objects fall under which concepts if we know which propositions are true. Evidently, the present discussion can be carried out either in terms of sentences and predicates or in terms of propositions and concepts.

Next, we can show that the reduction of holding to truth has a converse: if we know which predicates hold of which objects, then we can determine which sentences are true. At the level of natural language we can say

> The sentence A is true if and only if the predicate $\ulcorner$equals 1 whenever $A\urcorner$ does not hold of 0.

(The alternative version "the sentence A is not true if and only if the predicate $\ulcorner$equals 1 whenever $A\urcorner$ holds of 0" may be easier to parse.) The same thing can be accomplished more easily for formal languages by saying

A is true if and only if $\ulcorner A \wedge (x = 0) \urcorner$ holds of 0,

since $\ulcorner A \wedge (x = 0) \urcorner$ is a formula with one free variable and therefore qualifies as a formal language predicate. (The natural language translation of this formula is simpler than the expression I used above; I leave it to the reader to decide whether it counts as a predicate.) Thus, knowing which predicates hold of 0 is enough to tell us which sentences are true.

The grand conclusion we can draw is that knowing which predicates hold of which objects is equivalent to knowing which sentences are true. Like truth and the entailment relation (Section 1.4), the two are mutually recoverable.

At a formal level, holding is to formulas with one free variable what truth is to sentences. Thus the basic function of a holding relation is, similarly to a truth predicate, to allow us to ascribe concepts indirectly. Having this ability enables us to say things we could not otherwise say. For instance, we could describe an algorithm that generates a sequence of formulas, each with one free variable, and then say that the nth formula in the sequence holds of the natural number n. For a less contrived example of what we can do, consider the following:

> Let C be a concept. Suppose that 0 falls under C, and that for any natural number n, if n falls under C then so does $n + 1$. Then every natural number falls under C.

This is the *principle of induction*. It is an assertion about all concepts, and we need to use the notion of falling under to express it. The principle is usually stated only for sets, but as I suggested in the last section and will explain in detail in Chapter 6, the category of concepts is strictly broader than the category of sets — and this is even true in the natural number setting. (Every concept expressible in the language of arithmetic corresponds to a set, but we cannot say this of all concepts restricted to N.) So the full strength of the induction principle is only manifested in the version given above.

Comments about use and mention similar to those I made in Section 1.3 regarding truth can also be made here. To mention a predicate is to refer to it as a symbolic string, and to use it is to include it as a functioning component of an assertion. The idea is that enclosing a predicate in quotes converts use into mention, and applying the holding relation converts mention into use.

So truth and holding are intimately related. There is even a version of the T-scheme for holding, namely,

For all x, "$\mathcal{A}$" holds of x if and only if x $\mathcal{A}$,

where "$\mathcal{A}$" can be replaced by any predicate. For instance, "is white" holds of x if and only if x is white. The formal language version of this would be

For all x, "$\mathcal{A}$" holds of x if and only if $\mathcal{A}[x]$,

where "$\mathcal{A}$" can be replaced by any formula with exactly one free variable x.

Truth and holding are intertranslatable, have similar functions, and enjoy analogous schematic characterizations. Unfortunately, but predictably, holding shares not only the desirable features of truth, but also its paradoxical aspect. The analog of the liar paradox for holding arises from the *Russell predicate* "does not hold of itself". This predicate generates the paradoxical sentence

"Does not hold of itself" does not hold of itself,

which is (a version of) *Russell's paradox*. This can also be seen as essentially a rewording of Quine's paradox discussed in Section 1.1. The falling under version involves the *Russell concept*

being a concept that does not fall under itself.

Evidently, if the Russell concept falls under itself then it does not fall under itself, and if it does not fall under itself then it does fall under itself.

Russell's paradox has generated a vast amount of commentary which includes a variety of ideas as to the source of the difficulty. It should be clear from the earlier material in this section that the problematic element in the versions of the paradox just mentioned is the notion of holding/falling under, in the same way that the problematic element in the liar paradox is the notion of truth.

In particular, attempting to block Russell's paradox by denying the legitimacy of the Russell predicate is like attempting to block the liar paradox by denying the legitimacy of the liar sentence, and will fail in the same way. As I pointed out in Section 1.1, if the liar sentence is not a meaningful sentence then it cannot be a true sentence, and this leads directly to a contradiction. Similarly, if the Russell predicate is not a genuine predicate then it certainly cannot hold of itself. But to say that it does not hold of itself is just to say that "does not hold of itself" does not hold of itself.

This implies that "does not hold of itself" does hold of itself, and we have reached a contradiction.

This brings out an important difference between the concept/predicate versions of Russell's paradox and the analogous set-theoretic paradox. The latter version of the paradox is based on the construction "the set of all sets which are not elements of themselves", and it is standardly, and correctly, resolved by explaining that this "set" is not a set but a proper class. Thus the accepted resolution of the set version of Russell's paradox is the one we just rejected for concepts: it denies the validity of the construction. There is no "Russell set". But this is not a complete solution to the problem because it only succeeds by means of retreating to a broader category of object. The Russell "set" is really a class, and this immediately raises the question of whether there is a class of all classes which are not members of themselves. Again, sets and classes will be the subject matter of Chapter 6, and we will return to this topic then. For now, I simply note that the standard treatment of the set version of Russell's paradox is not of any help against the concept/predicate version.

2.3   Interpreted languages

> *General languages. Models. Truth functions for modelled languages. Interpretations.*

I described some specific constructions of truth predicates in Chapter 1. Concepts and relations can be used to give a much more general construction.[6]

Our general definition of a formal language $\mathbb{L}$ follows Frege's approach. *Terms* are built up in the usual way from an infinite list of variables ("x", "y", "z", ...) and some set of *operation symbols*. Operation symbols can be *binary*, taking two arguments (like addition), *unary*, taking one argument (like successor), or *nullary*, taking no arguments. Nullary operation symbols may be regarded as constant symbols (like "0"). We could also allow operations that take more than two arguments, but in practice this is rarely needed.

In order to form *atomic formulas* we also need a repertoire of *proposition, concept*, and *relation symbols*. Any proposition symbol is an atomic formula, if C is a concept symbol and t is a term then $\ulcorner C[[t]] \urcorner$ is an atomic formula, and if R is a relation symbol and t_1 and t_2 are terms then $\ulcorner R[[t_1, t_2]] \urcorner$ is an atomic formula. For example, the symbols "⊤" and

"⊥" introduced in Section 1.4 are proposition symbols, and "=" from that section is a relation symbol (with the inessential difference that I wrote $\ulcorner t_1 = t_2 \urcorner$ there instead of $\ulcorner = [[t_1, t_2]] \urcorner$). Again, one could generalize to allow more than two arguments if one wanted.

Once we have atomic formulas, we define arbitrary formulas in the usual way, building them up from atomic formulas using logical connectives and quantifiers. As before, a sentence is a formula with no free variables. This characterizes the general structure of an arbitrary formal language $\mathbb{L}$.

It does not make sense to talk about truth until we have some way of giving sentences meanings. This can be done quite naturally in the context of set theory. We describe a *model* for $\mathbb{L}$ by specifying the following data:

- a nonempty set M over which the variables are supposed to range
- interpretations of the operation symbols, as follows:
  - for each constant symbol, an element of M
  - for each unary operation symbol, a function from M to M
  - for each binary operation symbol, a function from M^2 to M
- interpretations of the proposition, concept, and relation symbols, as follows:
  - for each proposition symbol, a choice of either 1 or 0
  - for each concept symbol, a subset of M
  - for each relation symbol, a subset of M^2.

For instance, in the case of arithmetic the successor symbol is interpreted as the function $n \mapsto n + 1$ from $\mathbb{N}$ to $\mathbb{N}$ and the equality symbol is interpreted as the set $\{(n, n) : n \in \mathbb{N}\} \subset \mathbb{N}^2$.

In order to express negation, we can include a proposition symbol "⊥" ("bottom") and require that its interpretation be 0. Then $\ulcorner \neg A \urcorner$ could be taken as an abbreviation for $\ulcorner A \to \bot \urcorner$. Alternatively, the role of "⊥" could be played by any sentence which is false in the intended interpretation, such as $\ulcorner \hat{0} = \hat{1} \urcorner$ in the setting of arithmetic.

I will say that $\mathbb{L}$ is *modelled* if it is equipped with a set-theoretic interpretation of the above form. Given a modelled language, we can proceed to define a truth function τ. First, we can use the interpretations of the operation symbols in a straightforward way to define an interpretation in M of each constant term, i.e., each term that contains no variables. In order to follow the procedure we used for arithmetic, we must assume that every element of M is so represented. This can always be arranged by temporarily augmenting the language with a family of constant symbols, one for each element of M; then after we have defined τ we can restrict it to the sentences of the original language.

We define τ on atomic sentences as follows. If P is a proposition symbol then $\tau(P)$ is the value (either 1 or 0) that was chosen to be the interpretation of P; if C is a concept symbol and t is constant term then $\tau(C[[t]])$ is 1 or 0 depending on whether the interpretation of t is, or is not, an element of the interpretation of C, and if R is a relation symbol and t_1 and t_2 are constant terms then $\tau(R[[t_1, t_2]])$ is 1 or 0 depending on whether the pair of interpretations of t_1 and t_2 does, or does not, belong to the interpretation of R.

The definition of τ on more complex sentences proceeds recursively, in the same way as in the arithmetical case. In particular, the value of τ on sentences of the form $\ulcorner(\forall x)A[x]\urcorner$ or $\ulcorner(\exists x)A[x]\urcorner$ is determined by its values on the simpler (with one fewer quantifier) formulas $A[t]$ where t ranges over the constant terms.

That completes the description of how a truth function can be defined for a modelled language. It shows that the constructions presented in Chapter 1 are part of a general pattern. Can we affirm, as we did there, all statements of the form $\ulcorner\tau(``A") = 1 \leftrightarrow A\urcorner$ with A ranging over the sentences of $\mathbb{L}$? This is not quite as straightforward as it was in Chapter 1, because there we were analyzing notions (the natural numbers, the natural numbers mod n) that were also available in the informal metalanguage in which we had constructed the truth function. Thus a statement like $\ulcorner\tau(``\hat{1} + \hat{2} = \hat{3}") = 1$ if and only if $1 + 2 = 3\urcorner$ made sense and could be proven. The situation is different now because we cannot assume the sentences of $\mathbb{L}$ are also sentences of the metalanguage in which the analysis is taking place. However, this is not a serious problem because having a model for $\mathbb{L}$ tells us what the sentences of $\mathbb{L}$ mean. So it is possible to define, for each sentence A of $\mathbb{L}$, a translation $\hat{A}$ in the metalanguage that talks about the model. For example, the translation of $C[[t]]$ could be $\ulcorner$the interpretation of t belongs to the interpretation of $C\urcorner$. I omit the details. After this translation is defined, we can informally prove, in the metalanguage, all statements of the form $\ulcorner\tau(``A") = 1 \leftrightarrow \hat{A}\urcorner$ with A ranging over the sentences of $\mathbb{L}$.

Just as in Chapter 1, we have no obvious way to simultaneously affirm all such equivalences, even though we can see that they are all provable. I repeat the familiar refrain that a truth predicate is the standard tool for affirming a scheme, and here we would need to use a truth predicate whose range of applicability is wider than that of the one we are trying to define. To summarize: the special constructions of truth predicates relating to arithmetic and arithmetic mod n that I presented in Chapter 1 generalize to arbitrary modelled languages, and the same difficulties about expressing the fact that they succeed as truth predicates reappear in this more general setting.

But we should not concentrate exclusively on the idea of a modelled language. For our purposes even this setting is too limiting. For instance, if we are to have the ability to talk about arbitrary sets in a formal language then we have to allow for the possibility that the variables could range over a proper class (in this case, the class of all sets) rather than a set, which means that the language could not be interpreted in terms of a model in the above sense. In particular, there is no modelled language in which we are able to affirm facts about all modelled languges.

Perhaps one could take the position that arbitrary sets can only be discussed informally, but I reject this idea. Any precise thought that is capable of being expressed informally should be able to find rigorous formal expression. "What can be said at all can be said clearly".[7]

Going one step further, if we are to have the ability to talk about arbitrary classes in a formal language then we have to allow for the possibility that the variables could range over something more general than a class — since there is no class of all classes. The natural conclusion is that the most general form of an *interpretation* would involve variables which refer to the objects that fall under some specified concept (such as: being a natural number, being a set, being a class, being a concept). At this level of generality we should allow proposition, concept, and relation symbols to be interpreted not in terms of sets but simply as propositions, concepts, and relations. For instance, if the variables of the language range over classes (that is, they refer to the objects falling under the concept of being a class) then the equality symbol should be interpreted as the relation of equality among classes. The point is that this relation does not correspond to a set or even to a class. I will use the term *interpreted language* for a language equipped with an interpretation in this most general sense.

It is not so clear that truth functions can be defined in settings this broad, especially if (as I suggested above and will elaborate on later) we cannot affirm the law of excluded middle for arbitrary concepts. Indeed, assuming that a truth function can be defined for any interpreted language leads to a version of the liar sentence which states that the truth function for the language in which it resides evaluates it to 0. ("The truth function for English takes the value 0 on this sentence.") If we can talk about the general concept of an interpreted language and the general concept of a truth function for such a language, then we can formulate such a sentence, and the only conclusion to be drawn is that the language in which that sentence is formulated cannot have a truth function.

2.4 Global truth

Notions which require truth. Truth as a global concept. Is pretheoretic truth coherent?

The Tarskian catastrophe has emerged as a major obstacle preventing us from developing a satisfactory theory of truth. Although the T-scheme appears to exactly capture our intuitive notion of truth, it cannot be used as a definition because we are not able to assert it without referencing truth. In the setting of formal languages, we have found ourselves perfectly capable of constructing truth predicates for a variety of object languages, but unable to express the fact that we have done this without assuming that the metalanguage in which we have been working already carries a more encompassing truth predicate.

In addition, we have learned that other key notions can be derived from, and can be used to derive, a notion of truth. We saw this in Section 1.4 for the entailment relation and in Section 2.2 for the holding relation. Thus, until truth is available, neither of these relations is available either.

Since the identity criterion for concepts — two predicates express the same concept if and only if they hold of the same objects — involves the holding relation, it is not clear that we can legitimately talk about concepts without already having a notion of truth. The identity criterion for relations (two formulas, each with two free variables, express the same relation if and only if they hold of the same pairs of objects) also presumes a notion of truth, as does the identity criterion for propositions, via the entailment relation (two sentences express the same proposition if and only if they imply each other). We are building up a repertoire of notions:

- propositions
- concepts
- relations
- truth predicates for interpreted languages

that are intuitively meaningful but that we find ourselves unable to define without invoking a global notion of truth.

All of these problems would be solved if we could assume that we enter the discussion equipped with a pretheoretic global notion of truth that may be clarified by further analysis but does not have to be created from scratch. Indeed, many authors do assume this and some are openly dismissive of the idea that a definition of truth is needed.[8] So we must ask whether a coherent global concept of truth is pretheoretically available.

Given that doubt has been cast on the notion of a concept, it is unclear that this is even a meaningful question. To anticipate, my ultimate conclusion will be that all the notions mentioned above — propositions, concepts, relations, and truth predicates — are indeed legitimate when understood correctly and can in fact be defined without invoking truth. But that discussion will come later. In the meantime, any essential use of any of these notions must be understood by the reader as being contingent on some sort of future justification.

We can start the discussion of global truth by considering the numerical case. There should not be any question that in the setting of numerical sentences "is evaluated to 1 by τ" is a meaningful predicate. Some numerical sentences are evaluated to 1 by τ and others are not, and (modulo generic reservations about concepts/properties, as just mentioned) we are perfectly within our rights to express this by saying that some numerical sentences have the property of being evaluated to 1 by τ and others do not. That merely reflects our conception of "properties" as arising from any legitimate predicate. At a formal level, the relevant predicate is the formula "$\tau(A) = 1$", with one free variable, "A", which is taken to range over all numerical sentences.

Similar comments apply to arithmetic mod n with respect to $\tilde{\tau}$, as well as to ordinary arithmetic with respect to τ, provided the definition of τ in the full setting of arithmetic is accepted as meaningful, or indeed to any modelled language subject to the same qualification. We can therefore simply define "is true" in these settings to mean "is evaluated to 1 by τ" (or by $\tilde{\tau}$). If it is objected that this fundamentally computational definition fails to capture the deep essence of truth, we only have to point out that once it is admitted that the true sentences are precisely those which τ evaluates to 1, by definition "is true" and "is evaluated to 1 by τ" have to express the same concept. A concept is completely determined by specifying which objects fall under it.

Thus, to the question "is pretheoretic truth a valid concept?" we can answer that our pretheoretic idea of truth can be made rigorous in a variety of special settings. However, our concern now is with truth globally. From a realist perspective, this is a question about whether it is meaningful to talk

about the general truth of arbitrary interpreted sentences. (The reference
to an interpretation is needed because the same sentence could be true in
one interpretation but not another.) But the crucial difference between
reasoning in a metalanguage $\mathbb{L}_1$ about truth in a specific object language
$\mathbb{L}_2$ and reasoning in $\mathbb{L}_1$ about truth in an arbitrary language is that in the
latter case one should be able to specialize and reason in $\mathbb{L}_1$ about truth in
$\mathbb{L}_1$. So the essential problem is to formulate a language which contains a
predicate that captures our informal notion of truth for that very language.
This version of the problem should also be accessible to the nominalist.

There are a variety of reasons for suspecting either that this problem
ought to be solvable in some form or that it might be essentially unsolvable.
One piece of evidence is our ability to define truth predicates for modelled
languages. This may be grounds for limited optimism, as it shows that there
are many settings in which our informal idea of truth can be crystallized into
a definite concept. But these grounds are fairly weak since the problematic
aspect of self-reference is absent in these cases.

Another argument is that we certainly seem to have an informal ability
to talk about truth globally, which would indicate that there is such a thing
as global truth. But since we also have an informal ability to derive the liar
paradox, this argument does not seem so persuasive. It may be reasonable
to generally expect that notions which appear in ordinary language and
seem to be cogent should ultimately be capable of being made precise.
That may be a good starting point for theoretical analysis. But ordinary
language is not infallible, and one should be especially skeptical if the notion
in question is, in its naive usage, trivially inconsistent.

However, the argument from informal usage is not so easily dismissed
when we consider the strong intuition we had in Sections 1.4 to 1.6 that
we had a right to affirm various forms of the T-scheme, given that this is
apparently only to be done using a global notion of truth. Along the same
lines, going back to the comments I made at the beginning of this section,
any intuition we have for propositions, concepts, and relations as globally
meaningful notions could also be taken as support for a global concept of
truth in some form, to the extent that the meaningfulness of these notions
hinges on a notion of truth.

Realism specifically toward propositions could provide another reason
for believing in the validity of a global notion of truth, because "once one
admits that there are propositions, one is hard pressed to deny that there is
a property of precisely the true ones".[9] This is certainly right if one understands, as we do, "propositions" to be sentences modulo truth equivalence,

but if that is all that is meant then there is little reason to admit that there are such things as propositions before satisfying ourselves that we have a notion of truth. The argument has more force if "propositions" are taken in the sense of "meanings" or something of that nature. But in that case, admitting that there are such things as propositions is a much bolder prospect. Again, ordinary language is not infallible. The mere fact that we sometimes speak in a way that implies, or seems to imply, that there are such things as meanings may be evidential, but it is hardly conclusive.

The basic idea that truth describes a correspondence between language and reality could at first seem to support the idea of global truth. A sentence in an interpreted language *says something* about reality and it is true if what it says matches what is *really there*. This picture seems perfectly universal, at first. But on closer inspection it really only fits those cases where languages describe limited parts of reality, such as arithmetic. Languages which are meant to apply more globally face the obstacle that they could be part of the reality they are intended to describe. Of course, it is just this circular aspect which makes truth so slippery. When a sentence *says something* about itself it may be difficult to unravel just what that something is, if what it says could depend on what it says. In such cases our sense of the simple concreteness of truth loses its conviction. So the idea of truth as correspondence does not necessarily favor the reality of global truth.

The T-scheme can be a source of doubts about the reality of truth. One has a sense that it says everything about truth that needs to be said, which goes against the idea that truth has some deep, undefinable essence. It seems to be nothing more than a grammatical tool for disquotation which is assumed to take the form of a predicate. If that is all there is to it, then what can our naive grasp of truth consist in, other than an informal acquaintance with the T-scheme? If it then turns out that there is a circularity problem which formally prevents us from using the T-scheme to define truth, the best explanation could be that there cannot be any such predicate after all. From this perspective the real surprise would be that in the limited setting of a modelled language there always is a genuine concept that satisfies the T-scheme — though one is still at a loss to say what this "satisfaction" comes to without invoking truth. At any rate, the meagerness of the T-scheme seems to have led commentators more to question "whether truth has some sort of underlying nature"[10] than to doubt that it is a well-defined property.

Nothing said so far is conclusive. The decisive consideration that settles the question of whether we could have a coherent pretheoretic global concept of truth characterized by the T-scheme comes in the form of a theorem of Tarski. His *undefinability theorem* states that any interpreted formal language which meets a minimum standard of expressiveness cannot contain a truth predicate for itself.[11] (We will see the arithmetical version of this result in Theorem 3.8.) Informally, the basis of this theorem is the simple observation that the liar paradox renders a global concept of truth impossible. If we had a coherent pretheoretic global concept of truth then we could formulate a sentence (or a proposition) that denied of itself that it fell under this concept, and then we would get a contradiction from the T-scheme. What more needs to be said?

I think this argument is definitive. Usually when one runs into a contradiction that clear-cut one accepts the verdict. Perhaps the finality of the undefinability theorem is not fully appreciated. Or perhaps the liar paradox inspires less respect than it should because familiarity has bred contempt. Everyone has heard of the liar paradox, multiple solutions have been proposed, surely one of them will pan out; meanwhile we need not regard it as a fatal problem — this could be the thinking.

Indeed, maybe some resolution of the liar paradox will pan out. But let us take a moment to consider the limitations logic imposes on any possible solution. These limitations are actually quite severe because of the poverty of assumptions that go into the paradox. On the face of it there are only two ways to block the contradiction. Either one has to deny some essential law of logic, a desperate move which I will not consider further, or one has to reject an instance of the T-scheme — specifically, the instance where "A" is replaced by the liar sentence L. (Recall that L is the sentence "L is not true".)

What would denying the T-scheme for the liar sentence entail? If we retain the law of excluded middle then we have to accept that L is true or L is not true. Then in order for the equivalence

"L is not true" is true if and only if L is not true

to fail, one side would have to be true and the other false. Either we would have to affirm the liar sentence but deny that the liar sentence is true, or else we would have to affirm that the liar sentence is true but deny the liar sentence. Neither alternative is compatible with our naive idea of truth, by any stretch of the imagination.

One could avoid this dichotomy by rejecting the law of excluded middle, but recall that that alone does not block the paradox (see Section 1.1). We would *still* have to deny, or at least refuse to accept, an instance of the T-scheme.

The question here is not whether a workable notion of truth could be obtained by dropping some instances of the T-scheme. What we are asking now is whether we have a coherent pretheoretic notion of truth. And the answer to this question is that, in the presence of the law of excluded middle, no modification of the presumed properties of truth which blocks the liar paradox is remotely compatible with our naive idea of truth. If our pretheoretic notion of global truth is to be made into a precise concept we must face the outrageous requirement that this must be a concept for which we are able to affirm neither the law of excluded middle nor all instances of the T-scheme.

2.5 Defining truth

Hanging questions. Minimalism. Infinitely long sentences. Substitutional quantification.

We are moving toward the following conclusions with regard to the "two camps" mentioned in Section 1.1. In a variety of local settings, the second camp is right: truth is a real, well-defined concept. This means that for a variety of object languages, one is able to externally define a predicate which satisfies the T-scheme, in some sense of "satisfies" which we have yet to explain. But the first camp is right globally. Insofar as truth is characterized by the T-scheme, there is no meaningful global notion of truth, and correspondingly no language that contains a predicate which captures our informal idea of truth for that very language. (Or at least, no such language which meets some minimum standard of expressiveness. There could be trivial examples.) The most we can ask for in the self-applicative context is some sort of technical construction of a predicate that has some of the properties we expect of truth but not others.

This leaves a number of crucial questions unanswered. What does it mean to be a truth predicate for an interpreted language — that is, how can we make sense of the idea of the T-scheme being true without invoking truth? What could account for our intuitive sense that there is a global notion of truth? How can we properly define propositions, concepts, and relations without invoking truth? Can we somehow recover global versions of the entailment and holding relations?

Perhaps we were too hasty in abandoning the idea of global truth. There have been many proposals of different ways it might be rehabilitated; we had better look at a few of them. In this section we will consider a few attempts to define truth, and in the next section we will discuss some variant theories of truth.

Minimalism. According to the "minimalist" theory of truth it should be possible to use the T-scheme, or rather the T-scheme minus a few of its more pathological instances, to define truth from scratch.[12] The omissions would be chosen in a way that ensures the consistency of the defined notion but leaves intact all the instances of the T-scheme that we actually care about.

Since we cannot use a single sentence to assert a scheme without invoking truth, on this approach we do not attempt to do so. Instead, we simply regard all the instances of the T-scheme, minus omissions, as collectively defining truth. Mathematicians are familiar with the notion of an *axiom scheme*, where one includes alongside the ordinary axioms of some formal system a template that is used to generate an infinite list of additional axioms. There is no single statement that simultaneously affirms all the axioms in the list, but nonetheless, each individual axiom is available to be used. We could treat the T-scheme in this way.

In the last section I argued that denying any instance of the T-scheme does violence to our intuitive idea of truth. The argument there was that one cannot claim that we "just know" what truth is without needing a definition, and then have the actual nature of truth turn out to be radically different from our pretheoretic understanding. But this is not a problem for minimalists because their proposal does not depend on our having a coherent pretheoretic notion of truth. We do not need any prior acquaintance with truth to grasp the T-scheme as a scheme.

The real problem with the minimalist proposal is that it gives us no deductive power. The substantive value of having a truth predicate is that it enables one to affirm schematic assertions. If the truth predicate is itself introduced schematically, it loses this function.

Whenever one uses an axiom scheme, there is a danger of confusing the scheme with the single statement that all the axioms it generates are true. This one-sentence version of the principle expressed by the scheme is typically more powerful than the scheme itself, as it captures a global aspect which is missing from the individual instances of the scheme. In the case of minimalism, although we can grasp the general structure of the T-scheme, we do not grasp the scheme itself as a single principle, only as an infinite family of independent principles. Therefore any argument we make can explicitly invoke only finitely many instances of the T-scheme, and hence could also be made using a notion of truth that is only defined for finitely many sentences. But as we already discussed in Section 1.7, any finite set of instances of the T-scheme carries no content whatever. It is only when applied to infinitely many sentences that truth ascriptions display their true power. Merely having all instances of the T-scheme individually available gives us no ability to reason about the truth of any infinite family of sentences, and thereby deprives us of the main fruits of the notion of truth.[13]

For instance, suppose we want to argue that every sentence that is formally derivable in some axiomatic system S is true. How would we prove this? The obvious method would be to argue that every axiom of S is true and that the rules of inference preserve truth. But recognizing that the rules of inference preserve truth in general is not possible on the basis of finitely many instances of the T-scheme, which is all a minimalistically acceptable argument would allow us. One has to grasp global properties of truth, which one cannot do on the basis of a schematic definition.

Or suppose we want to prove that $\ulcorner A \to A \urcorner$ is true for every sentence A. For any given sentence A we can prove the sentence $\ulcorner A \to A \urcorner$ (this is a standard derivation; see Section 3.1), and then using an instance of the T-scheme we can prove that $\ulcorner A \to A \urcorner$ is true for this particular sentence A. We could go on to prove the same thing for finitely many other sentences, but that is the limit of our ability. The law that says $\ulcorner A \to A \urcorner$ is true for all A is *not a consequence* of the T-scheme. We know this because it is a general fact that any logical consequence of a set of axioms must be a consequence of some finite subset, and it is clear that no finite set of instances of the T-scheme can imply the global law. Since the minimalist proposal is to use the T-scheme, as a scheme, to define truth, this means that the law that says $\ulcorner A \to A \urcorner$ is true for every A is not a feature of minimalistic truth.

Horwich considers a propositional version of this problem (but for him, the identity criterion for propositions is intertranslatability, so his "propositions" are different from ours). He suggests that it could be overcome by supposing that "there is a truth-preserving rule of inference that will take us from a set of premises attributing to each proposition some property, F, to the conclusion that all propositions have F".[14] In the present example, F would be the property of $P \to P$ being true, as a property of the proposition P. But how could a minimalist know to apply such a rule? On the face of it, we would need to be dealing in infinitely long derivations in order to reach a point where we could draw a conclusion from an infinite set of premises. The only way to get there in a finite number of steps would be by somehow simultaneously grasping the truth of all the premises. But that is just the sort of thing that the minimalist would already have needed a rule of this type to accomplish.

A likely source of confusion is the fact that once we have understood the structure of the T-scheme, we can see globally that for any P there is a proof that $P \to P$ is true: essentially the same proof works in every case, so that one has what we could call a *proof scheme*. If an infinite rule of inference were available as a proof technique, we could then see that this infinite family of proofs can feed into the infinite inference rule, producing a single, infinitely long proof of the statement that $P \to P$ is true for all P. So we do know that there is an (infinitely long) proof of this statement. But that is where we have to stop, tantalizingly close to our goal. As I noted earlier, minimalism does not give us the global fact that everything provable is true, so we cannot take the final step and conclude that $P \to P$ is true for all P.

It would be futile to try to find a way to take this final step. We already know that minimalism does not give us the resources we need.

Infinitely long sentences. In Section 1.4 I briefly raised the possibility of using an infinitely long sentence, (†), to affirm the fact that τ provides us with a truth predicate. For that matter, we might consider using (†) to define τ. But we rejected (†) on the grounds that we needed a truth predicate to assert infinitely long sentences. Our reasoning was based on the implicit assumption that the metalanguage we are using does not itself contain sentences which are infinitely long.

But why not suppose that we work in a metalanguage that allows infinitely long sentences? This would apparently give us the ability to define a truth predicate for the object language. The obvious objection, that we are physically unable to speak or write such sentences, might be overcome by setting up a system for abbreviating (some) infinite expressions in a finite way. For instance, the literal expression (†), with two conjuncts followed by an ellipsis, could be understood as abbreviating a genuinely infinite conjunction. In such a way we might give ourselves the ability to reason using infinitely long expressions despite not being able to pronounce them in full. A more elegant technique would be to allow schematic assertions in the metalanguage, and to adopt the convention that any such assertion is to be interpreted as the conjunction of all its instances, with its schematic variables replaced in all possible ways by sentences of the object language.[15]

The problem with this idea is that it gives rise to an arms race between object language and metalanguage. Yes, once a metalanguage with infinitely long expressions is available, we can define truth predicates for languages that involve only finitely long expressions. But the metalanguage is not of that type, so we cannot define a truth predicate for the metalanguage, or even say what it means to be one, and thus there is now a catastrophe at the metalanguage level. This is a problem because we have the same intuitive sense as before that we do know what it means to be a truth predicate for the metalanguage, and moreover, we have the same ability to handle the catastrophe by going to a language with even longer infinite expressions.

This process can never stabilize — we can never formulate a metalanguage within which it is possible to form a version of (†) that includes every sentence of that very metalanguage. This is because the infinite conjunction that is (†) could, if it were part of the metalanguage, itself appear within sentences of the metalanguage, and would therefore have to be contained within some of its own conjuncts.

(In case the reader gets any ideas about using non-well-founded languages that would allow sentences to nest inside themselves, note that non-well-foundedness generates easy paradoxes. For instance,

$$(0 = 0) \land \neg((0 = 0) \land \neg((0 = 0) \land \cdots))$$

is effectively a liar sentence, without even any mention of truth. Non-well-founded sentences are problematic because there is no recursive procedure to assign them truth values or otherwise give them meaning.)

Thus, if we fix a metalanguage once and for all, there is a limit to the family of object languages for which we can say what constitutes a truth predicate. We do apparently have the ability to find, for any object language $\mathbb{L}_2$, a metalanguage $\mathbb{L}_1$ in which the notion of a truth predicate for $\mathbb{L}_2$ can be defined — but even this relatively modest assertion is problematic because there does not seem to be any language in which it can be made. (If $\mathbb{L}$ were such a language, set $\mathbb{L}_2 = \mathbb{L}$.) Ultimately, allowing infinitely long sentences merely delays the catastrophe.

Substitutional quantification. Another idea for defining truth is to modify the sense of the quantifiers "for all" and "there exists" to allow for quantification over schematic variables. Instead of *objectual* quantification, where the quantified variable ranges over a family of objects and serves as a variable name for those objects, we could employ *substitutional* quantification, where the quantified variable serves as a variable member of some grammatical category, such as predicates or sentences. Then we could express the T-scheme as a sentence:

For every sentence $\mathcal{A}$, "$\mathcal{A}$" is true if and only if $\mathcal{A}$.

Frege thought we could use quantifiers in this way, but that seems related to his belief that all grammatical forms, not just noun phrases, perform a referential function (see Section 1.2). If "snow is white" were the name of something, and asserting that snow is white were a matter of saying the name of that thing, then this kind of quantification might make sense.[16] But of course sentences are not names and this kind of quantification is simply ungrammatical. When we say "for every number n ..." we are committing ourselves to what follows when n is 1, or n is 2, or n is 3, etc. But when we say "for every sentence A ..." this cannot be specialized to the case when A is snow is white because "when A is snow is white" is not a cogent expression. "When A is 'snow is white'" is a meaningful phrase, but this shows that in what follows A must be replaced not by a sentence but by the name for a sentence. We can explain the problem as a use/mention conflict: the variable in a quantifying phrase has to be ordinary, not schematic, because it has to represent a mention, not a use. But in substitutional quantification later appearances of that variable need to represent a use.[17]

Suppose we agree that ordinary language does not allow us to quantify over schematic variables, and that we have no clear understanding of such quantification[18] — or even that such quantification is "obviously senseless"[19] — might we not solve the problem of substitutional quantification "by legislation"[20]? A number of authors have felt that we could. We should not feel obligated to follow the dictates of ordinary grammatical usage; if we can overcome a difficulty by altering the way we take language to work, maybe we should.

On the other hand, such an endeavor should not be undertaken lightly. We must be careful to ensure that our alterations make sense. And here we run into a difficulty that several authors have noted,[21] namely, that if we are going to use substitutional quantification to formulate the T-scheme as a sentence and thereby define truth, then we cannot use truth in our explanation of substitutional quantification. So the natural explanation in terms of truth conditions, via equivalences like

"$(\forall \mathcal{A})(\mathcal{A} \to \mathcal{A})$" is true if and only if $\ulcorner A \to A \urcorner$ is true for every sentence A, is disallowed.

Maybe we just know what substitutional quantifiers mean, without having to reduce them to anything more primitive? ("Explanations come to an end somewhere".[22]) But just as it does for truth (Section 2.4) the "we just know what it means" defense fails definitively for substitutional quantification because the naive, pretheoretic version of what we are supposed to "just know" is inconsistent. If we could schematically quantify over sentences then we could define truth globally via a quantified T-scheme, and then we could formulate a liar sentence which would engender a contradiction.

The reason for this phenomenon is the circularity inherent in sentences which involve quantification over all sentences. There is a use/mention distinction to be made here. If we are just considering the sentences of some language as syntactic objects, then no problem arises. The syntactic objects exist and can be talked about independently of any semantic interpretation we impose on them. But if we treat them as semantically meaningful then a vicious circle can arise when we try to make sense of a sentence whose meaning is taken to hinge of the meanings of all sentences, including its own.

For substitutional quantification to even have a chance, it has to be restricted in some way which is not naively obvious. I consider this a decisive refutation of the suggestion that we can take the meaning of substitutional quantification as naively understood.

One way in which substitutional quantification might be rehabilitated is by restricting the variables to range over some previously understood object language. This would avoid the circularity problem we just discussed. Then, e.g., "$(\forall \mathcal{A})(\mathcal{A} \rightarrow \mathcal{A})$" could be viewed as being synonymous with the (infinite) conjunction of the sentences $\ulcorner A \rightarrow A \urcorner$ with A ranging over the sentences of the object language, and "$(\exists \mathcal{A})(\mathcal{A} \rightarrow \mathcal{A})$" as being synonymous with their disjunction. There is no semantic circularity here since we already know what all the sentences of the form $\ulcorner A \rightarrow A \urcorner$ mean. Under this interpretation, the function of the substitutional quantifiers would simply be to abbreviate infinite conjunctions and disjunctions, so that we are now effectively back in the setting of infinitely long expressions (and the suggestion about using schematic variables made in that paragraph is simply a quantifier-free, and thus less expressive, version of what we are doing here). Thus, substitutional quantification could indeed be used in a metalanguage to define truth in an object language, but there would be no universal metalanguage in which truth for any language could be defined

— for exactly the same reason this could not be done with infinitely long
sentences.

2.6 The revenge problem

The pattern of revenge. Thoughts which cannot be expressed.
Vague truth. Revised truth. Situational truth. Grounded truth.
Determinate truth.

We continue our review of attempts to legitimize a global concept of
truth. Since the liar paradox is the core obstacle to be overcome, we can
frame the discussion around it. Indeed, if the liar paradox could be convinc-
ingly dissolved then we might not have to say anything more to rehabilitate
our naive idea of truth. That alone might be enough to persuade us to take
truth as a primitive concept not in need of being defined in any simpler
terms. However, treatments of the liar paradox generally come in the setting of a general analysis of truth.

Attempts to resolve the liar paradox are usually susceptible to the *revenge problem*, where an account of truth that purports to deal with the
paradox "will rely on concepts ... which, if allowed into the object language, generate new paradoxes that cannot be dissolved by the account in
question".[23] Typically the way this works is that the theory holds that ascriptions of truth lack some character of finality, the exact nature of which
depends on the account. Thus, saying that a sentence is true does not mean
that it is *really* true, only that it is true in some limited way. Then the idea
is that the liar sentence is not *really* true, but it is still, or still could be,
true in a limited sense. In this manner the paradox is averted.

The revenge in this scenario comes from a "strengthened liar" sentence
that asserts of itself that it is not *really* true. Even with the new, subtler
conception of truth in hand, one feels as though a contradiction can now
be derived exactly as in the classical liar paradox.

Defenses against the strengthened liar sentence vary, but often seem
to come down to some version of either denying that it can be properly
formulated — it "is not expressible in any language"[24] — or accepting the
formulation but claiming that "it does not say what we think it does",[25]
which more or less comes to the same thing.

Responses of this sort tend to be hard to swallow because one was under the impression that the entire discussion up to that point had been
conducted informally in a metalanguage in which the rogue concept was

perfectly expressible and said exactly what we thought it did. But apparently not. This idea that there are thoughts which cannot be expressed actually has an impressive pedigree, with fairly explicit endorsements from both Frege ("By a kind of necessity of language, my expressions, taken literally, sometimes miss my thought"[26]) and Wittgenstein ("There are, indeed, things that cannot be put into words ... They are what is mystical"[27]). Still, it seems like a cop-out. I urge the reader to reject it and I repeat my earlier comment that any precise thought that is capable of being expressed informally should be able to find rigorous formal expression.

The following are a handful of attempts to resolve the liar paradox which conform in varying degrees to the preceding outline. Details can be found in the cited references; I limit myself to giving brief summaries.

McGee proposes that truth be considered a vague predicate.[28] Thus some sentences are definitely true, some are definitely not true, and some are "unsettled" — their truth value is undetermined by the rules of our language. Evidently the liar sentence falls in the last category. The next step on the way to a contradiction would be to argue that this shows that the liar sentence is not true (and therefore it is true, since that is what it says), but that argument is not supposed to be valid because unsettled sentences could, after all, turn out to be true. So a contradiction is avoided.

The strengthened liar sentence in this case is "this sentence is not definitely true". Assuming it is definitely true leads to a contradiction, so it cannot be definitely true. Thus we have proven that it is not definitely true, i.e., we have proven the strengthened liar sentence itself, and since "whatever we can prove rigorously is definitely true",[29] it follows that the strengthened liar is definitely true. And as we noted at the start, this entails a contradiction. So a paradox is still present.[30]

Gupta and Belnap regard truth as inherently circular and argue that circular definitions should be accepted as legitimate, with the warning that the truth values of sentences involving circular concepts could be unstable.[31] They picture a series of stages, with the truth values at a given stage being used to determine the truth values at the next stage. Thus at each stage one judges the liar sentence to be either true or false, but this judgement is always liable to be revised. In particular, deducing the falsehood of the liar sentence from its truth, or vice versa, is merely a matter of revising one's assessment of its truth value. "Categorical" sentences eventually settle on a fixed truth value and "pathological" sentences do not.

This "revision theory" bears some resemblance to McGee's vagueness. There the pathological sentences were considered not to have a definite truth value, or at least not yet, but now they are considered to have truth values which keep changing. The *really* true sentences are now those which are categorically true, i.e., which eventually settle on being true, and the strengthened liar sentence is the sentence that states of itself that it is not categorically true.

Assuming this strengthened liar sentence is categorically true leads to a contradiction, so it must not be categorically true. It is hard to see how this conclusion could be open to revision, so that the assertion that the strengthened liar sentence is not categorically true must itself be categorically true. But that is what the strengthened liar sentence asserts. Gupta and Belnap counter that accepting the concept of categoricalness into our language engenders a "higher-level notion of categoricalness"[32], and that there is indeed a whole hierarchy of categoricalness predicates. The suggestion seems to be that there is a general phenomenon of truth values either remaining in flux or eventually settling, yet we are not able to describe it. Every time we invent a categoricalness predicate which could be used to this purpose, we unwittingly generate a "higher-level" notion outside the language. Thus the general phenomenon *is not expressible in any language* and a term like "categorical" which tries to capture it *does not say what we think it does.*

According to Barwise and Etchemendy, the truth of a sentence can only be judged relative to a "situation" — that part of the world that the proposition expressed by the sentence is about.[33] Thus there is no single liar proposition, but rather one liar proposition for each situation. Again we see a similarity with previous accounts: now truth values can change not because we have revised our judgement of them, but because we have passed to a more inclusive situation and the sentence whose truth we are judging expresses a different proposition relative to the new situation. One might suppose that a new paradox would emerge from a sentence that denies its truth in all situations (or maybe all "actual" situations[34]), but it turns out that this sentence *does not say what we think it does*, because "we cannot in general make statements about the universe of all facts".[35] This formulation is unfortunate because it seems itself to be a statement about the universe of all facts, and it therefore apparently disqualifies itself. Presumably the intention is that we should understand the impossibility of universally applicable statements to be something *which cannot be expressed*. But rather than legitimize the sentiment, all this does is to bring out the absurdity of trying to say that some true proposition cannot be expressed.

All three of the preceding theories picture truth as "emerging" in some way: an undetermined sentence can become true, an initially false sentence can become eventually true, or a sentence that is false in one situation can become true in a broader situation. In this respect these theories resemble an influential account due to Kripke.[36] Kripke is quite explicit that his account is not intended to resolve the liar paradox.[37] Indeed, it is not so much a theory of truth as it is a theory of "grounded" truth.

Suppose we are given a modelled language $\mathbb{L}$ as in Section 2.3 and we augment it with a concept symbol meant to represent self-applicative truth. That is, in the expanded language $\mathbb{L}'$ we have the syntactic ability to talk about the truth of sentences of $\mathbb{L}'$, including sentences that themselves contain truth ascriptions. Let G be the smallest set of sentences of $\mathbb{L}'$ such that (1) every sentence of $\mathbb{L}$ belongs to G — so all of these sentences have well-defined truth values; (2) if A belongs to G then so does $\ulcorner A$ is true$\urcorner$, with the same truth value as A; and (3) if the truth value of a sentence A can be evaluated via truth tables given the truth values of sentences in G, then A belongs to G. Thus if A belongs to G and is true, then $\ulcorner A \vee B \urcorner$ belongs to G and is true, for any sentence B; if $A[t]$ belongs to G and is false, for some constant term t, then $\ulcorner (\forall x) A[x] \urcorner$ belongs to G and is false; and so on. The *grounded* sentences of $\mathbb{L}'$ are those which belong to G.

The intuition is this. When we try to evaluate the truth of a sentence that refers to the truth of other sentences, we could end up chasing down a chain of truth ascriptions, or even multiple chains, and in the case of the liar sentence, this turns into an infinite regress which never determines the truth value of the original sentence. But if we are lucky, all the chains eventually terminate, or at least enough of them do to provide us with sufficient information to settle the truth value of the sentence we started with. In that case the sentence is grounded.

It is tempting to declare ungrounded sentences to be meaningless. But if we were to decide that true sentences must be grounded, then the liar sentence would not be true, and this would engender a contradiction in the usual way. Thus we cannot identify truth with grounded truth.

Soames interprets Kripke's account as a theory of *determinate* truth.[38] Rather like McGee's "unsettled" liar sentence which is not definitely true but could still be true, according to Soames the liar sentence is not determinately true but could still be true. So of course the strengthened liar sentence "this sentence is not determinately true" cannot be handled by the theory. Once again, "the very activity of solving the paradox in a particular language provides the material for re-creating another version of it in a broader context".[39]

2.7   Second-order logic

Second-order variables. Expressing global second-order facts.

As I have tried to explain in this chapter, I think there is no hope for a global concept of classical truth as characterized by the T-scheme. The T-scheme is inconsistent, by the liar paradox, and if it is rendered consistent by omitting some pathological instances, it no longer corresponds to our naive idea of truth. This shows that our naive idea of truth is, to the extent it is characterized by the T-scheme, incoherent. We cannot simply use an abridged version of the T-scheme to define a variant truth concept because we have no way to assert it (Tarskian catastrophe), and employing it as a scheme deprives it of all its functionality (minimalism). Attempts to reengineer our naive idea of truth do not succeed (revenge problem).

On the positive side, we have a general method for defining a truth predicate for any modelled language. There is still a problem of saying what it means to be a truth predicate — this is something which apparently can be done in any particular case by using a metalanguage which either allows infinitely long expressions or employs substitutional quantifiers which range over the sentences of the object language. But the "apparently" qualifier is necesary, because at this point we do not know how to formulate a language in which we would be able to express the fact that a suitable metalanguage can always be found.

So there are still problems to address, but the general outline of the situation is becoming clear: there is no global concept of classical truth, but there are local concepts which apply to particular object languages.

What I want to say now is that the preceding picture is, besides being incomplete in the manner indicated, utterly worthless from the perspective of second-order logic. *Second-order logic* is the term we use for reasoning which involves not only *first-order variables* which range over the objects of interest (such as number or sets), but also *second-order variables* which range over other grammatical categories — things like sentences about those objects, their properties, or relations between them.

We have discussed and rejected the idea of quantifying over schematic variables in Section 2.5. We must now acknowledge that this has the unwanted consequence of destroying second-order logic. For instance, if we want to say that every sentence implies itself, we seem to have two choices.

We can try to express this either schematically with the formula "$\mathcal{A} \to \mathcal{A}$" or in terms of an ordinary variable with the sentence "for every sentence A, $\ulcorner A \to A \urcorner$ is true". But the first alternative fails to capture the global fact, because it is a scheme and only finitely many instances of it could be used in any finitely long argument. And the second alternative *also* fails to capture the global fact, because it relies on a truth predicate, which by what I said above would have to be a local truth predicate, because that is the only kind of truth predicate we have.

This situation is unacceptable because every sentence, in any interpreted language, *does* imply itself. To give a less trivial example, it really is true that for any concept C, if 0 falls under C and whenever n falls under C so does $n + 1$, then every natural number falls under C. But without a global falling under relation we cannot express this global fact. And by the equivalence between truth and falling under (Section 2.2), if we only have local truth predicates then we only have local falling under relations.

The received view on second-order logic is that it is to be understood in set-theoretic terms. Thus, we can say that for any set X, if 0 belongs to X and whenever n belongs to X so does $n + 1$, then every natural number belongs to X. This is fine as far as it goes, but it fails to capture the truth of induction for concepts that cannot be represented as sets.

To make the point differently, consider that it is true of any concept C that if $\emptyset$ falls under C then $\emptyset$ falls under C. This is not only true of concepts that can be realized as sets. It is also true of the concept of being a set, under which concept $\emptyset$ does fall, and of the concept of being a proper class, under which concept $\emptyset$ does not fall. Interpreting second-order logic in terms of sets or classes would leave out these important cases; the general fact is a fact about concepts. But again, absent a global falling under relation we cannot express this general fact.

One can easily generate any number of similar ideas which seem cogent but which apparently cannot be formally expressed. Given the emphasis I have placed on the principle that any precise thought can be stated formally, this is a problem. The problem will be solved in Sections 5.5 and 5.6.

Chapter 3

Deduction

3.1 Natural deduction

Deduction and truth. Reasoning under assumptions. Introduction and elimination rules. Minimal, intuitionistic, and classical logic. Soundness.

From here on I will mostly stop quoting and quasi-quoting formal expressions. Although there is a big difference between saying that n is a number and saying that "n" is a variable, the usual practice is to leave this distinction implicit, making a trade-off in favor of readability over precision.

The concern of this chapter is formal deduction as a syntactic procedure for identifying the true sentences in an interpreted language. It is purely syntactic in the sense that the rules are mechanical; one could easily write a computer program to check the validity of a putative derivation, or to generate legal derivations at will.

In order to say anything substantive about which sentences are true, we need to assume that some knowledge is initially available. This will take the form of a family of sentences which are initially taken to be true; these are the premises or *axioms* to be used in derivations. Ideally, we would like a systematic procedure for generating all logical consequences of a given set of axioms. That is, we want to be able to determine all the sentences whose truth is guaranteed by the truth of the axioms.

No doubt the reader will have noted the key role of truth in the preceding remarks. Indeed, one of the main themes of this chapter is the relation between derivability and truth. Consequently, everything we do here requires that we work in a setting in which a truth predicate is available. For this reason we shall make a standing assumption that we are considering a modelled language. As we saw in Section 2.3, a truth function τ can be defined for any modelled language, at least if one is willing to accept the infinitary aspect of the definition. (If not, languages with finite models, such as arithmetic mod n, are still available.)

In terms of the truth function τ, what we are aiming for is a system of deduction which has the property that, when supplied with a family of sentences on each of which τ takes the value 1, will only generate sentences on which τ takes the value 1. This formulation allows us to avoid references to truth in favor of references to τ, which should bring clarity to the discussion. It also allows us to postpone saying what it means for "being evaluated to 1 by τ" to count as a truth predicate, a problem that we have yet to solve. Dealing directly with τ lets us put such philosophical questions to the side for now.

Since we do not know how to define truth functions in the more general setting of interpreted languages in the sense of Section 2.3, nothing I say in this chapter should be assumed to be valid there. However, this warning can be qualified by the comment that the deduction rules will, in fact, all turn out to be valid in the more general setting, and it should be intuitively plausible that this is so. But clarifying the sense in which they are more generally valid will be a problem for later chapters. (See Section 4.5.)

There are a variety of deductive systems available. Let us start with Gentzen's *natural deduction*, which has the merit of closely mirroring informal reasoning.[1]

In this style of reasoning we allow arbitrary formulas, not just sentences, to appear in derivations. We then ask for the universal closures of the derived formulas to be true, where a *universal closure* of a formula A is a sentence of the form $\overline{A} = (\forall x_1) \cdots (\forall x_n)A$ such that every free variable of A appears in the list x_1, ..., x_n. (The notation $\overline{A}$ is therefore slightly ambiguous, but this is not important because it generally does not matter in which order the quantifiers appear, or whether any are extraneous.) In general, when we say that a formula is true, we mean that its universal closure is true — more precisely, all of its universal closures are true.

In natural deduction, each logical symbol is associated to two rules, one by which it can be introduced and one by which it can be eliminated. Thus the introduction rule for $\wedge$ says that given both A and B, one can derive $A \wedge B$, and the elimination rule for $\wedge$ says that given $A \wedge B$, one can derive either A or B (or both).

The key feature of natural deduction is that one is able to reason under temporary assumptions. Thus, at any point in a derivation one can introduce an assumption, after which we may treat that formula as if it has been derived. Then under certain conditions assumptions may be *discharged*, after which they are no longer being assumed. A simple example of this is the introduction rule for →: given a derivation of B under the assumption A, we can discharge A and derive $A \to B$. The idea is that deriving B from A shows us that $A \to B$ is true, and once we have shown this, we can stop assuming A. Note that other assumptions may have been active in the derivation of B, and they will still appear as assumptions in the resulting derivation of $A \to B$.

Thus, assumptions can be nested. Also, in some of the rules we proceed on the basis of two or more independent previous derivations, so that derivations can have a tree-like structure, with distinct lines merging as the argument proceeds. At points where derivations merge, all of the assumptions used in any of the incoming derivations are transmitted as premises of the outgoing derivation, minus anything discharged at that step.

I also need to mention one technical point. The substitution of a term for a variable in some formula is said to be *free* if no variable in any of the substituted occurrences of the term becomes quantified in the process. For example, in the formula $(\exists y)(x = y)$, the term $z + \hat{1}$ can be freely substituted for x but the term $y + \hat{1}$ cannot. In the following it should be understood that all substitutions are required to be free.

All derivations are assumed to be finite. Legal derivations are characterized by the following rules for introducing and eliminating logical symbols.

Introduction rules

(1) Having separately derived both A and B, derive $A \wedge B$.
(2) Having derived either A or B, derive $A \vee B$.
(3) Having derived B under the assumption A, optionally discharge A and derive $A \to B$.
(4) Having derived $A[x]$, derive $(\forall x)A[x]$. (But x must not appear freely in any currently active assumption.)
(5) Having derived $A[t]$, derive $(\exists x)A[x]$.

Elimination rules

(1) Having derived $A \wedge B$, derive either A or B.
(2) Having derived $A \vee B$, and having separately derived C both under the assumption A and under the assumption B, optionally discharge A and B and derive C.

(3) Having separately derived A and $A \to B$, derive B.

(4) Having derived $(\forall x)A[x]$, derive $A[t]$.

(5) Having derived $(\exists x)A[x]$, and having separately derived B under the assumption $A[x]$, optionally discharge $A[x]$ and derive B. (But x must not appear freely in any currently active assumption or in B.)

Note that all instructions to discharge an assumption are optional; at such points we are also free to continue reasoning under the assumption.

For example, given three sentences A, B, and C, a derivation of the sentence $((A \to B) \wedge (B \to C)) \to (A \to C)$ could go like this:

Assume $(A \to B) \wedge (B \to C)$.

> Derive $A \to B$ and $B \to C$.
> Assume A.
>
>> Derive B from A and $A \to B$.
>> Derive C from B and $B \to C$.
>
> Derive $A \to C$
> (discharging A).

Derive $((A \to B) \wedge (B \to C)) \to (A \to C)$
(discharging $(A \to B) \wedge (B \to C)$).

The rules do not mention negation; as before, we can take $\neg A$ to abbreviate $A \to \bot$, where in the general setting of a modelled language $\bot$ can be a propositional symbol that is not defined in terms of anything more primitive but which is required to satisfy $\tau(\bot) = 0$. (Or, as in the case of arithmetic, we could use a specific sentence such as $\hat{0} = \hat{1}$ in place of $\bot$.) How $\bot$ is treated depends on the nature of the reasoning one is engaged in. In *minimal logic* no special assumptions about $\bot$ are made. In *intuitionistic logic* we are allowed to assume $\bot \to A$ for any sentence A, and these assumptions are never discharged. In *classical logic* we are allowed to assume $\neg\neg A \to A$ for any sentence A, and these assumptions are never discharged. Depending on which form of logic we are using, we shall say that a formula is *miminally, intuitionistically,* or *classically* derivable from the set of premises. (These assumptions can also be presented as rules: given $\bot$, derive A, or given $\neg\neg A$, derive A.)

The law that characterizes intuitionistic reasoning, which says that any sentence follows from a falsehood, is called *ex falso quodlibet*, and the classical law that any sentence follows from its double negative is called *double negation elimination*. The ex falso law is also a feature of classical logic; it can be derived from double negation elimination, as can the law of excluded middle. These are both good exercises.

It is not hard to show that natural deduction has the desired property of preserving truth. Recall that $\overline{A}$ is any universal closure of A.

Theorem 3.1. *(Soundness of natural deduction) Let τ be the truth function for a modelled language and suppose the formula A is classically derivable from a set of sentences X. If $\tau(C) = 1$ for every C in X, then $\tau(\overline{A}) = 1$.*

Proof. It is easy to see that $\tau(\neg\neg C \to C) = 1$ for any sentence C. This is automatic. So instead of working with classical derivations, we can assume that X contains all instances of the double negation law and take the derivation of A to be minimal.

As in Section 2.3, we can suppose that every element of the model is represented by a constant term. Then for any function ϕ from the set of variables to the set of contant terms and any formula C, let C^ϕ be formula obtained from C by replacing its free variables with constant terms according to ϕ. Call such a function ϕ a *variable assignment*. Observe that $\tau(\overline{C}) = 1$ if and only if $\tau(C^\phi) = 1$ for every variable assignment ϕ. Therefore it will suffice to prove that for any set of formulas X, any variable assignment ϕ, and any formula A derivable using minimal logic from X, if $\tau(C) = 1$ for every $C \in X$ then $\tau(A^\phi) = 1$.

The proof goes by induction on the length of the derivation of A. It is just a matter of examining the final step of the derivation. For some of the rules this is quite trivial; for instance, if the final step is to derive $A \wedge B$ from A and B, we just observe that the induction hypothesis ensures that $\tau(A^\phi) = \tau(B^\phi) = 1$, and this immediately implies that $\tau(A^\phi \wedge B^\phi) = 1$. In other cases we have to allow ϕ or X to vary. For instance, if the final step is to derive $A \to B$ from a derivation of B under the assumption A, then we appeal to the induction hypothesis with $X \cup \{A^\phi\}$ in place of X. This tells us that if $\tau(A^\phi) = 1$ and $\tau(C) = 1$ for all $C \in X$ then $\tau(B^\phi) = 1$. Thus also $\tau(A^\phi \to B^\phi) = 1$. But if $\tau(A^\phi) = 0$ then trivially $\tau(A^\phi \to B^\phi) = 1$, so we can conclude that $\tau(A^\phi \to B^\phi) = 1$ only under the assumption that $\tau(C) = 1$ for all $C \in X$, as desired. I leave the remaining cases to the reader. $\qquad\square$

This theorem shows that formal derivations do what we want them to do in the setting of modelled languages. But at the same time it shows that this is not good enough, for the theorem itself evidently cannot be expressed in such a language; it talks about arbitrary modelled languages, which means that we need variables which can range over the class of all sets. So the proof of the soundness theorem involves deductive reasoning in a setting to which the theorem does not apply. This is an issue we will eventually have to address.

3.2   The completeness theorem

The model existence theorem. The completeness of natural deduction. Hilbert systems.

According to Theorem 3.1, natural deduction is *sound* — anything we can classically derive from true premises will also be true. The converse question is whether every logical consequence of some set of premises is classically derivable from them. Since a given set of premises could be true in more than one model, the right way to interpret this question is to take "logical consequence of X" to mean "a formula that is true in any model in which every sentence in X is true". The answer is yes; in this section I will sketch a proof of this statement. It is called the *completeness theorem* for natural deduction because it shows that natural deduction cannot be substantively strengthened. If a formula A is not derivable from X then there is a model in which every sentence in X is true but A is false.[2]

Say that a set of sentences X is *classically consistent* if $\perp$ is not classically derivable from X. The key step in the proof of the completeness theorem is the *model existence theorem* which states that every classically consistent set of sentences is true in some model.[3] We start with two lemmas.

Lemma 3.2. *Let X be a set of sentences in a language and let A be a sentence which is not classically derivable from X. Then the set $X \cup \{\neg A\}$ is classically consistent.*

Proof.   Suppose $X \cup \{\neg A\}$ were not classically consistent, i.e., suppose $\perp$ were classically derivable from $X \cup \{\neg A\}$. We must show how to convert such a derivation into a derivation of A from X. First, discharge $\neg A$ and derive $\neg A \to \perp$, i.e., $\neg\neg A$, from the premises X; then, reasoning classically, derive A from X. Since this contradicts the hypothesis that A was not derivable from X, we conclude that the set $X \cup \{\neg A\}$ had to have been classically consistent. $\square$

Lemma 3.3. *Let X be a classically consistent set of sentences in a language* *and suppose $A[x]$ is a formula with one free variable such that $(\exists x)A[x]$ is in X. If we introduce a new constant symbol a into the language, then the set $X \cup \{A[a]\}$ is classically consistent.*

Proof. Suppose $\perp$ were classically derivable from $X \cup \{A[a]\}$. Without loss of generality we can assume the variable x does not appear anywhere in this derivation; then replacing a with x throughout and appending a single use of the elimination rule for $\exists$ yields a derivation of $\perp$ from $X \cup \{(\exists x)A[x]\} = X$. Since we assumed X was classically consistent, we conclude that $\perp$ could not have been derivable from $X \cup \{A[a]\}$. $\quad\square$

Theorem 3.4. *(Model existence theorem) Let X be a classically consistent set of sentences in a language. Then there is a model with respect to which $\tau(C) = 1$ for every C in X.*

Proof. We start by normalizing X in two ways. First, we will need it to be maximal in the sense that for every sentence A of the language, either A or $\neg A$ belongs to X. Second, for every sentence of the form $(\exists x)A[x]$ in X we will need there to be a constant symbol a such that $A[a]$ belongs to X. These two conditions can be achieved in a series of stages. In the first stage, add sentences to X one by one, until nothing more can be added without creating an inconsistency. At this point the resulting set X' will be maximal, because any sentence A is either derivable from X', and hence could have been added, or else, by Lemma 3.2, its negation could have been added. In the second stage, add a constant symbol, say a, to the language for each sentence of the form $(\exists x)A[x]$ in X', and simultaneously add $A[a]$ to X'. This cannot introduce an inconsistency, by Lemma 3.3. Call the resulting set of sentences X''. This completes the second stage. Since we have just added constants to the language, there are now some new sentences, so X'' may no longer be maximal. So we repeat the first stage to get a new set X''', and so on. Then $\widetilde{X} = X \cup X' \cup \cdots$ has the desired properties. Obviously, if we can find a model for $\widetilde{X}$ it will also be a model for X.

Define a model for $\widetilde{X}$ as follows. Let M be the set of constant terms of the enlarged language. The interpretation of any constant term is itself; this shows both how to interpret the constant symbols and how to interpret the operation symbols. Second, the interpretation of each proposition symbol P is either 1 or 0 depending on whether P does or does not belong to $\widetilde{X}$; the interpretation of each concept symbol C is the set of $t \in M$ such that $C[[t]]$ belongs to $\widetilde{X}$; and the interpretation of each relation symbol R is the set of pairs $(t_1, t_2) \in M^2$ such that $R[[t_1, t_2]]$ belongs to $\widetilde{X}$.

We must verify that every sentence in $\widetilde{X}$ is true in the model. We can do this by induction on the number of logical connectives and quantifiers in the sentence. For atomic sentences it is immediate from the way we defined the interpretation. For a sentence of the form $A \wedge B$ in $\widetilde{X}$, maximality implies that both A and B must belong to $\widetilde{X}$ (since clearly neither $\neg A$ nor $\neg B$ can belong to $\widetilde{X}$, lest it be inconsistent). So the induction hypothesis ensures that $\tau(A) = \tau(B) = 1$, and therefore $\tau(A \wedge B) = 1$ by the definition of τ. For sentences of the form $A \vee B$ in $\widetilde{X}$, maximality implies that at least one of A and B must belong to $\widetilde{X}$ (otherwise both $\neg A$ and $\neg B$ belong to $\widetilde{X}$, which creates an inconsistency with $A \vee B$), and for sentences of the form $A \to B$ in $\widetilde{X}$, maximality implies that at least one of $\neg A$ and B belongs to $\widetilde{X}$ (otherwise both A and $\neg B$ belong to $\widetilde{X}$, which creates an inconsistency with $A \to B$). In either case we again get the desired conclusion from the induction hypothesis and the definition of τ. For sentences of the form $(\forall x)A[x]$ in $\widetilde{X}$, maximality implies that $A[t]$ belongs to $\widetilde{X}$ for every constant term t, since these are all derivable from $(\forall x)A[x]$, so that $\tau(A[t]) = 1$ for all such terms by the induction hypothesis, and therefore $\tau((\forall x)A[x]) = 1$. Finally, for sentences of the form $(\exists x)A[x]$ in $\widetilde{X}$, the construction of $\widetilde{X}$ ensures that $A[a]$ belongs to $\widetilde{X}$ for some constant a, which again allows us to deduce from the induction hypothesis that $\tau(A[a]) = 1$ and hence that $\tau((\exists x)A[x]) = 1$. This completes the proof.   □

The completeness theorem is a straightforward consequence of the model existence theorem together with Lemma 3.2.

Corollary 3.5. *(Completeness of natural deduction) Let X be a set of sentences in a language and let A be a formula which is not classically derivable from X. Then there is a model with respect to which $\tau(C) = 1$ for every C in X but $\tau(\overline{A}) = 0$.*

Proof.   Since A is not classically derivable from X, neither is $\overline{A}$, so Lemma 3.2 shows that the set $X \cup \{\neg \overline{A}\}$ is classically consistent. The model existence theorem then guarantees that there is a model with respect to which $\tau(C) = 1$ for every $C \in X$ and $\tau(\neg \overline{A}) = 1$, so that $\tau(\overline{A}) = 0$.   □

I have formulated the soundness and completeness theorems in terms of natural deduction because this is the most intuitive deductive system. However, it can also be somewhat awkward in its use of temporary assumptions and formulas containing free variables. Therefore, it will be convenient to have an alternative deductive system available in which derivations are presented linearly, with no temporary assumptions, and only involving sentences. The form of deduction I will describe now is called a *Hilbert system*.

In Hilbert systems we adopt as axioms all sentences that have one of the following forms:

(1) $A \wedge B \to A$
(2) $A \wedge B \to B$
(3) $A \to (B \to A \wedge B)$
(4) $A \to A \vee B$
(5) $B \to A \vee B$
(6) $(A \to C) \to ((B \to C) \to (A \vee B \to C))$
(7) $A \to (B \to A)$
(8) $(A \to (B \to C)) \to ((A \to B) \to (A \to C))$
(9) $\bot \to A$
(10) $(\forall x)A[x] \to A[t]$
(11) $A[t] \to (\exists x)A[x]$
(12) $(\forall x)(A \to B[x]) \to (A \to (\forall y)B[y])$
(13) $(\exists x)(B[x] \to A) \to ((\exists y)B[y] \to A)$

where in (10) and (11) the substitution of t for x must be free and in (12) and (13) the variable x must not occur freely in A and the variable y either equals x or does not occur freely in B. I will call the above *Hilbert axioms*. A formal derivation from a set of sentences X is then a finite sequence of sentences, each of which either belongs to X, or is a Hilbert axiom, or follows from two previous sentences in the list by the *universal modus ponens* rule which permits the deduction of $\overline{B}$ from $\overline{A}$ and $\overline{A \to B}$. That is, infer any universal closure of B from any universal closures of A and $A \to B$.

The Hilbert axioms presented above characterize intuitionistic logic. For minimal or classical logic the ex falso law can be omitted or the law of excluded middle can be included, respectively. Thus we can characterize minimal, intuitionistic, and classical reasoning in terms of Hilbert systems as well as in terms of natural deduction. The results are equivalent; I omit the proof.[4]

Theorem 3.6. *Any sentence derivable in a minimal, intuitionistic, or classical Hilbert system can also be derived in the corresponding system of natural deduction, and conversely.*

3.3   Peano arithmetic

The Peano axioms. Their intuitive content. Finite decidability.

Theorem 3.1 and Corollary 3.5 give us a complete picture of classical
derivability in the setting of modelled languages: anything derivable from a
set of axioms will be true in any model that makes those axioms true, and
anything not derivable from them will be false in some model that makes
them true. If we are interested in determining the true sentences in some
particular modelled language, then, our focus should be on finding a good
set of axioms, ideally one from which all true sentences of the modelled
language can be derived. In the next few sections we will see how this
works out in the case of arithmetic.

I introduced the language of arithmetic in Section 1.5. The standard
axioms are the universal closures of the following:[5]

(1) $x = x$
(2) $x = y \to y = x$
(3) $(x = y \land y = z) \to x = z$
(4) $\neg(0 = x')$
(5) $x = y \leftrightarrow x' = y'$
(6) $x + 0 = x$
(7) $x + y' = (x + y)'$
(8) $x \cdot 0 = 0$
(9) $x \cdot y' = x \cdot y + x$
(10) $\mathcal{A}[0] \land (\forall x)(\mathcal{A}[x] \to \mathcal{A}[x']) \to (\forall x)\mathcal{A}[x]$.

These are the *Peano axioms*. They are all single axioms except the induction scheme (10), in which $\mathcal{A}$ can be replaced by any formula (possibly
with free variables other than x). *Peano arithmetic (PA)* is the system of
deduction which employs the Peano axioms and classical logic. In general
we use the term *formal system* to refer to a set of axioms in a given formal
language together with a choice of logic. If S is a formal system, we say
that a formula is *derivable in S* if there is a derivation of the formula using
the appropriate form of logic from the axioms of S.

The Peano axioms have the same character of simplicity and inevitabil-
ity that we saw in the rules of natural deduction. The first three axioms
merely describe the nature of equality and do not relate specifically to arithmetic. Axioms (4) and (5) tell us that the elements of the sequence 0, 0',
0'', ... are all distinct — one never returns to 0 because 0 is not a successor,

and one never returns to any other earlier point since two different numbers cannot have the same successor. The next four axioms then specify, recursively, how addition and multiplication are defined on the elements of the number sequence.

The only remaining question is whether the sequence of successors of 0 exhausts the model. It is not so clear how to ensure this; we want to say that every number is obtained by applying the successor operation to 0 finitely many times, but the notion of "finitely many times" cannot be formally expressed without invoking the notion of a natural number that we are trying to characterize. The best we can do is the induction scheme, which intuitively says that properties of 0 that are transmitted to successors are transmitted to every number. However, it only says this of properties expressible in the language.

How much of ordinary informal reasoning about numbers does Peano arithmetic capture? The answer is, a great deal. Of course, "ordinary informal reasoning about numbers" is not a precise notion, and probably the best way to get a really clear understanding of what is possible is by going through the tedious process of building up the body of material that constitutes basic number theory within PA. It may help to point out what kinds of reasoning it does not accomodate, at least not straightforwardly. This includes any reasoning that involves literally infinite structures such as manifolds or vector spaces, or any kind of infinitary reasoning beyond what is captured by the induction scheme. In particular, PA is not well equipped to formalize reasoning that explicitly invokes the truth function for arithmetic, since its definition involves infinitely long computations.

On the other hand, there is a sense in which PA is not limited to reasoning about numbers. It can also be used to formalize elementary reasoning about other kinds of finite structures, to the extent that these structures are characterized by a finite amount of information which can be encoded number-theoretically. For example, given a symbolic alphabet, we can assign each character a binary code, and then replacing each character in a finite symbolic string with its binary code results in a single (probably very large) binary number. Indeed, it is in this sort of way that computers normally handle symbolic data. So by encoding such data numerically, we gain the ability to reason about it within PA.

The language of computation can also help to clarify what is arithmetically expressible. I mentioned earlier that concepts like "being even" and "being prime" can be expressed by arithmetical formulas. In fact something much more general can be said: any *finitely decidable property* is so expressible. That is, given any property P of numbers, if we can write a computer program that for any given number can determine in finitely many steps whether that number has P, then there is a formula $A[x]$ in the language of arithmetic with one free variable x such that

$$A[\hat{n}] \text{ is true} \quad \leftrightarrow \quad n \text{ has the property } P.$$

For example, since the primality of a given number can be mechanically tested, this tells us that there has to be a formula that expresses the concept of being prime.

More generally, for any k, any finitely decidable property of k-tuples of numbers can be expressed by an arithmetical formula with k free variables. So, for instance, we know that "n is a power of m" can be expressed arithmetically because we can mechanically check, for any m and n, whether n is a power of m. (Assuming $m \geq 2$, evaluate its powers in increasing order until n is either reached or exceeded.) The emphasis is on *finite* decidability — no claim is made about the expressibility of properties that can be tested with infinitely long computations. Indeed, we will soon see examples of such properties that are not arithmetically expressible.

The statement that finitely decidable properties can be arithmetically expressed should help give an idea of the scope of what can be discussed in the language of arithmeetic. I omit its proof. The notion of a finitely decidable property is informal, but it can be made precise via the *Church-Turing thesis*, which states that the notion of a mechanical algorithm is captured by any of several equivalent standard models of computation.[6] Once one has a specific programming language or model of computation in mind, the claim that all properties of numbers decidable in that model are arithmetically expressible is a rigorous assertion and can be proven.

As to what kinds of reasoning about these properties are supported by PA, the general idea is that this includes all "elementary" reasoning. Making this assertion more precise would take us too far afield, so I will work at an informal level and ask the reader to accept that the development could be made rigorous with more work.

3.4   The first incompleteness theorem

Gödel numbering. Finitely decidable axioms. The Gödel sentence. The first incompleteness theorem. Nonstandard models. Tarski's undefinability theorem.

The ability to reason about symbolic data within PA opens up the possibility of analyzing the notion of derivability in arbitrary formal systems, and in particular in PA itself. As I mentioned above, symbolic strings can be encoded numerically. Fix some encoding of the language of arithmetic, so that every formula has a numerical code or *Gödel number*. I will denote the numeral of the Gödel number of A by $\langle A \rangle$. This notation will be useful because we are going to reason about Gödel numbers formally within PA.

We have to assume that the encoding procedure is sufficiently reasonable that we can perform the sort of computations on Gödel numbers that we intuitively expect to be able to. For instance, we can tell by looking at a formula whether it has the overall form $A \wedge B$, and if it does we can read off the formulas A and B. So given the Gödel number for a formula we would like there to be a mechanical algorithm for making this determination, and for evaluating the codes of the constituent subformulas. If Gödel numbers were assigned arbitrarily then this might not be possible, but under any straightforward encoding convention it should not be a problem. I will suppress technical details and simply assume that a reasonable encoding has been chosen.

Going one step further, formal derivations in PA are characterized by a finite amount of information, so they can be assigned Gödel numbers too. Just as before, we must assume that the encoding is reasonable enough that it allows us to perform simple compuations. For example, given two numbers x and y it should be possible to mechanically check whether y encodes a derivation in PA of a sentence encoded by x. Our discussion of finite decidability in the last section then tells us that there is an arithmetical formula $\text{Der}[x, y]$ which is true if and only if there is a classical derivation D from the Peano axioms of a sentence A such that x is the Gödel number of A and y is the Gödel number of D. Then we can define $\mathbb{D}[x]$ to be the formula $(\exists y)\text{Der}[x, y]$, and this formula will hold of just those values of x which are the Gödel numbers of sentences derivable in PA. Thus, in PA we have the remarkable ability to reason about which sentences are derivable in PA. (Again, we will need to assume that the formula $\text{Der}[x, y]$ is chosen reasonably. I will not keep repeating this comment; assume that it applies universally.)

What is important about PA for our ability to arithmetize derivability in this way is that the Peano axioms are themselves *finitely decidable* in the sense that one can mechanically check whether a given formula is one of the axioms, and be assured of getting an answer after a finite computation. (An older, less apt term for this property is *recursive*.) Thus we can generalize the notation defined above to other axiomatic systems, even systems that use other languages. In this chapter we will be particularly interested in *extensions* of PA, formal systems which include all the Peano axioms and employ either the language of arithmetic or an augmentation of that language. Thus, if S is any finitely decidable extension of PA, then there is an arithmetical formula $\text{Der}_S[x, y]$ which expresses that there is a derivation D in S of a sentence A such that x is the Gödel number of A and y is the Gödel number of D. (As the language varies, the Gödel numbering scheme will have to vary as well, but we shall suppress this dependence in our notation.) We define $\mathbb{D}_S[x]$ to be the formula $(\exists y)\text{Der}_S[x, y]$; this expresses that x is the Gödel number of a sentence that is derivable in S. When S = PA I will continue to just write "$\text{Der}[x, y]$" and "$\mathbb{D}[x]$".

The next ingredient we need is numerical substitution. If $A[x]$ is a formula with one free variable x and n is a number, we have used the notation $A[\hat{n}]$ for the formula obtained by replacing every free occurrence of x with the numeral $\hat{n}$. Since this is a mechanical procedure, it should be possible to mechanically verify that the Gödel numbers of $A[x]$ and $A[\hat{n}]$ are related in the stated manner. Thus there is an arithmetical formula $\text{Sub}[x, y, z]$ which expresses that x is the Gödel number of a formula with one free variable and z is the Gödel number of the sentence obtained by replacing every free appearance of that variable with the numeral $\hat{y}$.

What we are moving toward is a sentence that asserts of itself that it is not derivable in PA. We just have to figure out how to get the formula to refer to itself. Now the intuition for the operation of numerical substitution is that it combines a predicate (a formula with one free variable) with a noun phrase (a numeral) to get a sentence: it is a sentence-building operation. We can use it to follow the model of Quine's paradoxical sentence from Section 1.1 and rig a syntactic construction that makes a sentence refer to itself. The trick is to consider a sentence obtained from a formula with one free variable by numerically substituting into that formula its own Gödel number. Thus let $G[x]$ be the formula

$$(\forall y)(\text{Sub}[x, x, y] \rightarrow \neg\mathbb{D}[y]).$$

If n is the Gödel number of a formula $A[z]$ with one free variable z, then $G[\hat{n}]$ says there is no derivation in PA of the sentence $A[\hat{n}]$ obtained by substituting the numeral $\hat{n}$ into that formula. Finally, let g be the Gödel number of the formula $G[x]$ itself and consider the *Gödel sentence* $G[\hat{g}]$. In words, this sentence says there is no derivation in PA of the sentence obtained by substituting the numerical value of g into G; that is, it says that there is no derivation of itself. This is a formal version of Quine's sentence mentioned in Section 1.1, with derivability replacing truth.

Neither the Gödel sentence nor its negation can be derived in PA. More generally, let S be a finitely decidable formal system in the language of arithmetic that is a *sound* extension of PA — every axiom of PA is also an axiom of S, and all additional axioms of S are true statements of arithmetic. Then we can draw an analogous conclusion about S.[7]

Theorem 3.7. *(Gödel's first incompleteness theorem) The Gödel sentence is true but it is not derivable in PA. If S is an extension of PA that uses the same language and is finitely decidable and sound, then there is an arithmetical sentence that is true but not derivable in S.*

Proof. Suppose $\tau(G[\hat{g}]) = 0$. Then there must be a value of y for which $\tau(\text{Sub}[\hat{g}, \hat{g}, \hat{y}]) = 1$ but $\tau(\neg\mathbb{D}[\hat{y}]) = 0$, i.e., $\tau(\mathbb{D}[\hat{y}]) = 1$. But the only value of y which makes $\text{Sub}[\hat{g}, \hat{g}, y]$ true is the Gödel number of the sentence obtained by substituting the numerical value of g into G — that is, the Gödel number of $G[\hat{g}]$. Thus, there must be a derivation in PA of $G[\hat{g}]$.

Granting that we know $\tau(A) = 1$ for every Peano axiom A, it follows from Theorem 3.1 that $\tau(G[\hat{g}]) = 1$. But we had assumed that $\tau(G[\hat{g}]) = 0$. This contradiction shows that $\tau(G[\hat{g}])$ must be 1, i.e., the Gödel sentence must be true. But, taking $y = \hat{g}$, it then follows that there is no derivation in PA of the sentence $G[\hat{g}]$.

An identical argument proves the same result for any finitely decidable, sound extension of PA. We just have to use the formula $\mathbb{D}_S[x]$ in place of $\mathbb{D}[x]$ when constructing the Gödel sentence. □

The first part of this theorem tells us that Peano arithmetic is incomplete — there are true arithmetical sentences which cannot be derived in PA. It is interesting to note that according to Lemma 3.2 this implies that there are false arithmetical sentences which can consistently be added to the Peano axioms, and by Theorem 3.4 there must then be models of PA in which some arithmetically false statements are true. What would such a *nonstandard model* look like? When I introduced the Peano axioms in Section 3.3 I pointed out that they ensure the existence of a sequence of elements $0, 0', 0'', \ldots$, but they do not obviously guarantee that this sequence exhausts the model. We now see that PA indeed does have nonstandard models which are not exhausted by this sequence.

What is more, the incompleteness theorem shows that this defect is not specific to PA. Any finitely decidable extension by true formulas is vulnerable to the same construction of a true but underivable sentence.

If the Gödel sentence is not derivable in PA, how did we show that it is true? We did this using the truth function τ, which is not available in PA. The infinitary aspect of τ, the fact that it is computable only in a generalized sense that allows infinitely long computations, excludes it from the realm of elementary number theory. Indeed, we can show that the predicate "is the Gödel number of a true arithmetical sentence" cannot be expressed in the language of arithmetic.[8] This is a special case of Tarski's undefinability theorem mentioned in Section 2.4.

Theorem 3.8. *(Tarski's undefinability theorem for arithmetic) There is no arithmetical formula with one free variable that holds of precisely the Gödel numbers of true arithmetical sentences.*

Proof.   Suppose there were such a formula $\mathbb{T}[x]$. Let $G'[x]$ be the formula

$$(\forall y)(\text{Sub}[x, x, y] \rightarrow \neg \mathbb{T}[y]),$$

let g' be the Gödel number of G', and consider the sentence $G'[\hat{g}']$. There is precisely one value of y for which $\text{Sub}[\hat{g}', \hat{g}', y]$ is true, namely the Gödel number of $G'[\hat{g}']$. So $G'[\hat{g}']$ is equivalent to the assertion $\neg \mathbb{T}[\hat{n}]$ where n is the Gödel number of the sentence $G'[\hat{g}']$, and therefore $G'[\hat{g}']$ and $\mathbb{T}[\hat{n}]$ must have opposite truth values. But the hypothesis on $\mathbb{T}$ says that $\tau(G'[\hat{g}']) = \tau(\mathbb{T}[\hat{n}])$. This contradiction establishes the theorem. $\square$

3.5   The second incompleteness theorem

> Con(S). *The second incompleteness theorem. The system* PA*. *Löb's theorem.*

Theorem 3.7, in the form in which I have stated it, cannot be derived in PA. Since it references the truth of the Gödel sentence, according to Theorem 3.8 it cannot even be expressed in the language of arithmetic. However, by framing the result in terms of consistency rather than soundness, we can get a version which is derivable in PA.

Let S be a finitely decidable extension of PA. We now allow the possibility that the language could be augmented (so we do not automatically have a notion of soundness for S). Then S is consistent if $\perp$ is not derivable in S, and this can be expressed arithmetically by the sentence $\text{Con}(S) \equiv \neg \mathbb{D}_S[\langle \perp \rangle]$. Thus, although we cannot explicitly reason in PA about the concept of arithmetical truth, we can reason about the consistency of PA and of finitely decidable extensions of PA.

The following theorem can be formulated in the language of arithmetic and derived in PA. In order for its proof to make sense, one has to get used to the idea that elementary reasoning about formal systems can be formalized in PA. In particular, elementary facts about what is derivable in PA should themselves be derivable in PA. We just have to avoid reference to concepts like soundness that cannot be expressed arithmetically.

Theorem 3.9. *(Gödel's second incompleteness theorem) Let S be a finitely decidable extension of PA. If S is consistent then Con(S) is not derivable in S.*

Proof. Let G_S be the Gödel sentence associated to S. Thus, we can derive, in PA, the equivalence $G_S \leftrightarrow \neg\mathbb{D}_S[\langle G_S \rangle]$, and it follows that we can also derive this in S. Now suppose G_S could be derived in S. Then we could infer $\neg\mathbb{D}_S[\langle G_S \rangle]$. However, some number y would encode a derivation in S of G_S, and this property of y would be an elementary combinatorial fact that could also be verified in PA, and hence in S. So in S we could derive both $\mathbb{D}_S[\langle G_S \rangle]$ and $\neg\mathbb{D}_S[\langle G_S \rangle]$. We conclude that if S is consistent then G_S cannot be derived in S.

The preceding argument can be formalized in PA, and hence in S. So we can derive the implication

$$\mathrm{Con}(S) \to \neg\mathbb{D}_S[\langle G_S \rangle]$$

in S. Thus, if Con(S) could be derived in S then so could $\neg\mathbb{D}_S[\langle G_S \rangle]$, and hence so could G_S. But we just saw that this implies S is inconsistent. So if S is consistent then Con(S) cannot be derivable in S. $\square$

This version of the incompleteness result is more illuminating about the nature of the incompleteness phenomenon. The Gödel sentence is somewhat contrived, and one may be unsure what to make of the fact that Peano arithmetic fails to derive it. The failure of PA to derive its own consistency is more meaningful, especially in the context provided by Theorem 3.9, which says that all finitely decidable, consistent extensions of PA share this deficiency. Thus, we could strengthen PA by adding Con(PA) as an axiom to get a stronger system PA' in which we could, trivially, derive that PA is consistent. But according to the second incompleteness theorem, PA' will not be capable of deriving its own consistency.

Once we grasp the structure of the set of natural numbers, it is easy to see that the Peano axioms are all true, i.e., they are all evaluated to 1 by the truth function τ — provided one accepts the infinitely long computations which go into the definition of τ. So we know from Theorem 3.1 that PA is consistent, and hence that all the axioms of PA$'$ are true. Thus, by Theorem 3.1 again, PA$'$ is consistent. The process could be iterated by now adding Con(PA$'$) as an axiom to get an even stronger system PA$''$. But by Theorem 3.9 this system is still not able to derive its own consistency. That is a point we can never reach.

In fact, it is possible to formulate a sentence which effectively affirms the consistency of the system PA* obtained from PA by adding that sentence itself as an axiom. (Thus, by the second incompleteness theorem the sentence must be false and PA* must be inconsistent.) To do this, let NSub$[x, y, z]$ be an arithmetical formula which expresses that x is the Gödel number of a formula with one free variable and z is the Gödel number of the *negation* of the sentence obtained by replacing every free appearance of that variable with the numeral $\hat{y}$. Then let $G^*[x]$ be the formula

$$(\forall y)(\text{NSub}[x, x, y] \to \neg \mathbb{D}[y]).$$

This formula expresses that if x is the Gödel number of a formula $A[z]$ with one free variable, then when the numerical value of x is substituted into $A[z]$ the result is a sentence whose negation is not derivable in PA. Then letting $\hat{g}^* = \langle G^*[x] \rangle$, we get a sentence $G^*[\hat{g}^*]$ which says of itself that its negation is not derivable in PA. Let PA* be the system obtained by adding the sentence $G^*[\hat{g}^*]$ to PA as an additional axiom.

The sentence $G^*[\hat{g}^*]$ has a circular quality, but it appears to only be circular in the same harmless way as the sentence "this sentence is true". However, it is false, as a consequence of Theorem 3.9 and the fact that the consistency of PA* can be derived within PA*. For $G^*[\hat{g}^*]$ implies Con(PA*) by Lemma 3.2, and this implication can be derived in PA. Since $G^*[\hat{g}^*]$ is trivially derivable in PA*, this means that Con(PA*) can be derived in PA*.

Similar reasoning can be used to prove an elegant generalization of the second incompleteness theorem due to Löb. The second incompleteness theorem says that if Con(S) is derivable in S then S is inconsistent, where Con(S) is the assertion that $\bot$ is not derivable in S. We could also describe Con(S) as the assertion $\neg \mathbb{D}_S[\langle \bot \rangle]$, i.e., the assertion that $\mathbb{D}_S[\langle \bot \rangle]$ implies $\bot$. So an equivalent version of the second incompleteness theorem says that if

$$\mathbb{D}_S[\langle \bot \rangle] \to \bot$$

(i.e., Con(S)) is derivable in S then

$$\bot$$

is derivable in S. Löb's theorem generalizes this to any sentence in place of $\perp$.

Theorem 3.10. *(Löb's theorem) Let* S *be a finitely decidable extension of* PA *and let* A *be a sentence in the language of* S. *If* $\mathbb{D}_S[\langle A \rangle] \to A$ *is derivable in* S *then* A *is derivable in* S.

Proof. To get a contradiction, suppose that $\mathbb{D}_S[\langle A \rangle] \to A$ is derivable in S but A is not. By Lemma 3.2, the system $S' = S + \neg A$ obtained from S by adding $\neg A$ as an axiom is consistent. Now $\neg A$ is trivially derivable in S', and since every axiom of S is also an axiom of S', we also have by hypothesis that $\mathbb{D}_S[\langle A \rangle] \to A$ is derivable in S'. Putting these together yields that $\neg\mathbb{D}_S[\langle A \rangle]$ can be derived in S'.

Reasoning in S', we can now invoke Lemma 3.2 (more precisely, the special case of this lemma when X is finitely decidable) to conclude, just as above, that $S + \neg A$ is consistent. That is, S' derives its own consistency. According to Theorem 3.9, it follows that S' is actually inconsistent. We have reached a contradiction, and this proves the theorem. $\square$

The significance of this result is that it shows that S has no nontrivial ability to affirm its own soundness. A statement of the form $\mathbb{D}_S[\langle A \rangle] \to A$ can be thought of as saying that if A is derivable then it is true, so all such statements collectively affirm that S is sound. But according to Löb's theorem, the only instances of this scheme that are derivable in S are those for which A is derivable in S. In other words, we only know that the derivability of A implies its truth in those cases where we already know that A is true.

3.6 Arithmetic with truth

The axioms of PA$^{\mathbb{T}}$. *Consistency and soundness of* PA. *The first incompleteness theorem for* PA.

Theorem 3.7 cannot be formalized in PA because it invokes the truth function for arithmetic and the fact that PA is sound. But we can create an environment in which this kind of reasoning can be performed by augmenting PA with a basic ability to handle arithmetical truth. I will call the resulting system PA$^{\mathbb{T}}$.

The language of PA$^{\mathbb{T}}$ is the language of arithmetic augmented by a single concept symbol $\mathbb{T}$. The formula $\mathbb{T}[x]$ is interpreted as expressing that x is the Gödel number of a true sentence of arithmetic, i.e., that x is the Gödel number of a sentence A with $\tau(A) = 1$. Here I refer to a Gödel numbering of the language of arithmetic, not the augmented language that includes $\mathbb{T}$. Thus $\mathbb{T}$ cannot be applied to sentences containing itself.

In order to put the axioms of $\mathrm{PA}^{\mathbb{T}}$ in as simple a form as possible, I introduce some notational conventions. First of all, if x is the Gödel number of A and y is the Gödel number of B, then we can mechanically calculate the Gödel number of $A \wedge B$ from x and y. Call this value $\mathrm{Conj}[x, y]$. Now the expression $\mathrm{Conj}[x, y]$ is not literally part of the language of arithmetic, but it can be treated as a natural abbreviation. For instance, if we wanted to assert that the value $\mathrm{Conj}[x, y]$ satisfies some formula $B[z]$, we could write this as $B[\mathrm{Conj}[x, y]]$, and take this expression to be an abbreviation of, say,

$$(\forall z)(A[x, y, z] \to B[z])$$

where $A[x, y, z]$ is an arithmetical formula which expresses that z is the Gödel number of the conjunction of two formulas whose Gödel numbers are x and y. In this way, we can informally introduce extra operation symbols into the language.

Secondly, we will want to make assertions which quantify over the Gödel numbers of sentences. For instance, we want to say that for any sentences A and B we have

$$\mathbb{T}[\langle A \wedge B \rangle] \leftrightarrow \mathbb{T}[\langle A \rangle] \wedge \mathbb{T}[\langle B \rangle]$$

— $A \wedge B$ is true if and only if A is true and B is true. In order to express this assertion in the language of arithmetic we could write something like

$$(\forall x)(\forall y)(\mathrm{Sent}[x] \wedge \mathrm{Sent}[y] \to (\mathbb{T}[\mathrm{Conj}[x, y]] \leftrightarrow \mathbb{T}[x] \wedge \mathbb{T}[y])),$$

where $\mathrm{Sent}[x]$ expresses that x is the Gödel number of a sentence and $\mathbb{T}[\mathrm{Conj}[x, y]]$ is understood in the manner indicated above. From here on I will freely use the simpler, less literal formulations, with the understanding that they are merely abbreviations for more complicated expressions.

The logic of $\mathrm{PA}^{\mathbb{T}}$ is classical. Its axioms consist of the Peano axioms, including the induction scheme for all formulas of the augmented language — so we can make induction arguments on formulas that involve $\mathbb{T}$ — together with six additional axioms which characterize the truth predicate. The six extra axioms are (universal closures of)

(1) $\mathbb{T}[\langle t_1 = t_2 \rangle] \leftrightarrow e[\langle t_1 \rangle] = e[\langle t_2 \rangle]$

(2) $\mathbb{T}[\langle A \wedge B \rangle] \leftrightarrow \mathbb{T}[\langle A \rangle] \wedge \mathbb{T}[\langle B \rangle]$

(3) $\mathbb{T}[\langle A \vee B \rangle] \leftrightarrow \mathbb{T}[\langle A \rangle] \vee \mathbb{T}[\langle B \rangle]$
(4) $\mathbb{T}[\langle A \rightarrow B \rangle] \leftrightarrow (\mathbb{T}[\langle A \rangle] \rightarrow \mathbb{T}[\langle B \rangle])$
(5) $\mathbb{T}[\langle (\forall x) A[x] \rangle] \leftrightarrow (\forall x) \mathbb{T}[\langle A[\hat{x}] \rangle]$
(6) $\mathbb{T}[\langle (\exists x) A[x] \rangle] \leftrightarrow (\exists x) \mathbb{T}[\langle A[\hat{x}] \rangle]$.

In axiom (1), the quantification is over all numerical terms t_1 and t_2, and e is an "evaluation" operation that takes the Gödel number of a numerical term to its numerical value. (I assume here a Gödel numbering of the numerical terms.) This operation can be expressed arithmetically because, of course, numerical terms can be evaluated by a simple mechanical procedure. The remaining axioms are self-explanatory. Thus, the axioms for $\mathbb{T}$ specify its behavior on atomic sentences and its interaction with the logical connectives and quantifiers. This is enough to give us a version of the T-scheme:

Theorem 3.11. *For any formula $A[x_1, \ldots, x_k]$ in the language of arithmetic with free variables $x_1, \ldots, x_k$, the universal closure of*

$$\mathbb{T}[\langle A[\hat{x}_1, \ldots, \hat{x}_k] \rangle] \leftrightarrow A[x_1, \ldots, x_k]$$

is derivable in $\mathrm{PA}^{\mathbb{T}}$.

Proof. The proof goes by induction on the complexity of A. In the atomic case, we have to show, in $\mathrm{PA}^{\mathbb{T}}$, that $\mathbb{T}[\langle \hat{t}_1 = \hat{t}_2 \rangle]$ and $t_1 = t_2$ are equivalent for an arbitrary choice of (not necessarily numerical) terms t_1 and t_2, where the notation $\hat{t}$ indicates a numerical term obtained by replacing free variables with numerals. By axiom (1) for $\mathbb{T}$, this reduces to showing that $e[\langle \hat{t}_1 \rangle] = e[\langle \hat{t}_2 \rangle]$ if and only if $t_1 = t_2$, which is done by induction on the complexity of the terms t_1 and t_2.

The remaining cases are straightforward consequences of the induction hypothesis and the axioms for $\mathbb{T}$. For instance, if we already have $\mathbb{T}[\langle A \rangle] \leftrightarrow A$ and $\mathbb{T}[\langle B \rangle] \leftrightarrow B$, then we can derive

$$\mathbb{T}[\langle A \wedge B \rangle] \quad \leftrightarrow \quad \mathbb{T}[\langle A \rangle] \wedge \mathbb{T}[\langle B \rangle] \quad \leftrightarrow \quad A \wedge B$$

using axiom (2). $\qquad \square$

The preceding proof illustrates the importance of being able to make induction arguments on sentences that contain $\mathbb{T}$. It is not just the axioms (1) — (6) that give us the ability to reason globally about truth, but those axioms in conjunction with the expanded induction scheme.

We can also verify the consistency and soundness of PA within $\mathrm{PA}^{\mathbb{T}}$. In the next result, $\mathbb{D}$ continues to refer to derivability in PA.

Theorem 3.12. *The sentence*

$$(\forall x)(\mathbb{D}[x] \to \mathbb{T}[x])$$

is derivable in PA^{T}.

Proof. We start by arguing in PA^{T} that $\mathbb{T}[\langle A \rangle]$ holds when A is any Peano axiom or any sentence of the form $\neg\neg B \to B$. We have this for the first nine Peano axioms by nine separate applications of the T-scheme (Theorem 3.11). For the induction scheme, it will suffice to show that for any formula $B[x]$ with one free variable we have

$$\mathbb{T}[\langle B[0] \wedge (\forall x)(B[x] \to B[x']) \to (\forall x)B[x]\rangle],$$

or equivalently,

$$\mathbb{T}[\langle B[\hat{0}]\rangle] \wedge (\forall x)(\mathbb{T}[\langle B[\hat{x}]\rangle] \to \mathbb{T}[\langle B[\hat{x}']\rangle]) \to (\forall x)\mathbb{T}[\langle B[\hat{x}]\rangle].$$

To do this, define an operation $f[m,n]$ such that $f[\langle B[x], n\rangle] = \langle B[\hat{n}]\rangle$; thus, if m is the Gödel number of a formula with one free variable then $f[m,n]$ is the unique number which satisfies $\mathrm{Sub}[m, n, f[m,n]]$. Then the desired statement is

$$\mathbb{T}[f[\hat{m}, \hat{0}]] \wedge (\forall x)(\mathbb{T}[f[\hat{m}, \hat{x}]] \to \mathbb{T}[f[\hat{m}, \hat{x}']]) \to (\forall x)\mathbb{T}[f[\hat{m}, \hat{x}]],$$

which is an instance of the induction scheme for PA^{T}. So $\mathbb{T}$ holds of all of the Peano axioms. Also, for any sentence B, using the T-scheme we have

$$\mathbb{T}[\langle B \to \bot\rangle] \leftrightarrow (\mathbb{T}[\langle B\rangle] \to \bot),$$

i.e., $\mathbb{T}[\langle \neg B\rangle] \leftrightarrow \neg\mathbb{T}[\langle B\rangle]$. So the law $\mathbb{T}[\langle \neg\neg B \to B\rangle]$ is equivalent to the statement $\neg\neg\mathbb{T}[\langle B\rangle] \to \mathbb{T}[\langle B\rangle]$ and can be derived from the sentence $(\forall x)(\neg\neg\mathbb{T}[x] \to \mathbb{T}[x])$, which is available because PA^{T} assumes classical logic.

The remainder of the proof is basically a formalization of the proof of Theorem 3.1. We must show that if $\mathbb{T}$ holds of finitely many sentences then it holds of the universal closure of any formula derivable from those sentences. This is done by a double induction, primarily on the length of the derivation and subordinately on the number of sentences in the premise. (As in Theorem 3.1, we are using natural deduction. The proof could also be formulated in terms of Hilbert systems by making an argument in PA^{T} that $\mathbb{T}$ holds of all instances of the Hilbert axiom schemes.) □

Corollary 3.13. *The sentence* $\mathrm{Con(PA)}$ *is derivable in* PA^{T}. *For every sentence A in the language of arithmetic, the sentence* $\mathbb{D}[\langle A\rangle] \to A$ *is derivable in* PA^{T}.

Proof. As we observed earlier, the first statement is equivalent to the special case of the second when A is $\bot$. The second statement follows immediately from Theorems 3.12 and 3.11. $\qquad\square$

This gives us the resources we need to prove the first incompleteness theorem within PA^{T}.

Corollary 3.14. *The sentences* $\mathbb{T}[\langle G[\hat{g}]\rangle]$ *and* $\neg\mathbb{D}[\langle G[\hat{g}]\rangle]$ *are both derivable in* PA^{T}.

Proof. We can run through the argument given in the proof of Theorem 3.7. Reasoning in PA^{T}, assume $\neg G[\hat{g}]$. We can then derive $\mathbb{D}[\langle G[\hat{g}]\rangle]$ (this inference can already be obtained in PA), and according to Corollary 3.13, we can also derive $\mathbb{D}[\langle G[\hat{g}]\rangle] \to G[\hat{g}]$. This yields $G[\hat{g}]$, which contradicts the assumption, so we must have $G[\hat{g}]$. Then $\mathbb{T}[\langle G[\hat{g}]\rangle]$ follows from Theorem 3.11, and $G[\hat{g}] \to \neg\mathbb{D}[\langle G[\hat{g}]\rangle]$ is already derivable in PA, so we also get $\neg\mathbb{D}[\langle G[\hat{g}]\rangle]$. $\qquad\square$

It is crucial to the preceding results that we be able to reason hypothetically about truth. If all we had was the T-scheme as expressed in Theorem 3.11 then, having derived any particular sentence A, we could go on to derive $\mathbb{T}[\langle A\rangle]$. But in the proof of Corollary 3.14, for instance, we do not actually have a derivation of $G[\hat{g}]$; rather, we argue that if we did then it would have to be true, using general properties of truth and the assumption that it follows from true premises by a finite chain of reasoning. In order to make this argument we need to know global laws about truth: for any A and B, if A and B are both true then so is $A \wedge B$, and so on.

Thus, what gives PA^{T} its power is the fact that we were able to characterize the truth predicate recursively using only finitely many axioms (the axioms (1) — (6) above).

3.7 Formal truth predicates

Formal truth predicates. Relative to an axiomatization, not an interpretation. Unable to affirm general laws. Problems for second-order logic.

Tarski uses the term "formal language" to mean what we would now call a formal system — a language together with a set of axioms and a choice of logic.[9] (In his work on truth Tarski assumed the use of classical logic.) This is just a matter of terminology, but it suggests a way of thinking that was probably more common before the incompleteness phenomenon was discovered. His assimilation of a choice of axioms into the notion of what constitutes a language suggests a pre-Gödelian confidence in the stability of those axioms, whereas the incompleteness theorems apparently show that we have an open-ended ability to strengthen any explicitly presented set of axioms for arithmetic, and thus that our grasp of the natural numbers cannot be encapsulated in any finitely decidable set of axioms.

This view of the primacy of formal systems informs Tarski's method of handling the Tarskian catastrophe I introduced in Section 1.3. Recall that the problem is how to say what constitutes a truth predicate for an interpreted language, as affirming the T-scheme would seem to require us to employ an even broader truth predicate than the one we are trying to define. Tarski's solution is to assume that the predicate in question has been defined in a metasystem — not a metalanguage, but a formal system whose language contains the language to which the predicate is applied — and to require that each instance of the T-scheme be derivable in that metasystem. Thus, we finesse the problem of how to say that each instance of the T-scheme is true by replacing "true" with "formally derivable". Since the latter is a purely syntactic notion, there is no vicious circle.

On this view, what counts as a truth predicate for arithmetic, say, varies depending on the metasystem one is using. Indeed, there can be cases where, according to one metasystem, some definition characterizes a truth predicate, but according to another, it does not (see below). This is related to the fact that different metasystems can disagree about which sentences are true. We have complete freedom in this regard: any sentence A whose negation is not a theorem of the target system S can consistently be added to S as an axiom, by Lemma 3.2; and then we can find a metasystem S' that defines a truth predicate for the augmented system $S + A$. This would also be a truth predicate for the original system S, and the statement that A is true would be derivable in the metasystem S'. If neither A nor its negation is derivable in S, this means that there will be metasystems according to which A is true and metasystems according to which $\neg A$ is true.

For example, in the case of Peano arithmetic there are metasystems with truth predicates for PA in which it is derivable that Con(PA) is true, and there are metasystems in which it is derivable that Con(PA) is not true.

This seems wrong because one feels that there is an intended interpretation of the language of arithmetic which is not captured by any finitely decidable axiomatization, and the truth of an arithmetical sentence should be defined relative to this interpretation, not relative to a formal system.

But this problem is not necessarily fatal; one could respond that our grasp of the intended interpretation of the language of arithmetic will be reflected in our choice of metasystem. Truth predicates appearing in other metasystems might correspond to alternative interpretations.

However, the characterization of truth predicates in terms of formal derivability suffers from a more severe defect, essentially the same problem facing minimalism (Section 2.5). Namely, it completely fails to capture the global nature of truth which is really the whole point of having a truth predicate. Here, as there, the attempted characterization is schematic; the metasystem has to separately prove each instance of the T-scheme, not a single statement (that one would not even know how to formulate) that affirms all instances simultaneously. Thus, while the predicate $\mathbb{T}[x]$ defined in Section 3.6 counts as a truth predicate for PA by virtue of Theorem 3.11, we could define a much weaker truth predicate for PA by simply taking all instances of the T-scheme as axioms. The latter truth definition has no deductive power, because any formal derivation could use only finitely many instances of the T-scheme, and predicates that satisfy any finitely family of instances of the T-scheme can already be defined within PA by using a finite version of the sentence (†) from Section 1.4.[10]

To bring home the complete plasticity of Tarski's characterization of truth, consider a predicate $\mathbb{T}'$ that is defined on Gödel numbers of arithmetical sentences by taking $\mathbb{T}'[\langle \mathrm{Con(PA)} \rangle]$ as an axiom, together with all instances of the T-scheme minus the single instance

$$\mathbb{T}'[\langle \mathrm{Con(PA)} \rangle] \leftrightarrow \mathrm{Con(PA)}.$$

If one then adds $\mathrm{Con(PA)}$ as an axiom, then $\mathbb{T}'$ counts as a truth predicate for PA — every instance of the T-scheme is derivable. But if one adds $\neg\mathrm{Con(PA)}$ as an axiom, then $\mathbb{T}'$ does not count as a truth predicate for PA, as one instance of the T-scheme fails.

But the situation is even worse than this. Tarski tells us that truth is to be understood in a purely formal way. It has no conceptual content — we are simply entitled, in some metasystems, to refer to some predicates as truth predicates. The whole point is that we do not talk about the T-scheme really being true in some universal sense; all that matters is whether every instance of the scheme can be derived in a given metasystem, and there is no absolute notion of some metasystems being more "correct" than others. Now suppose we want to affirm that $A \to A$ is true for every arithmetical sentence A. This would have to be done in a metasystem which contains a truth predicate $\mathbb{T}$ for PA, and the quantification over all sentences could be accomplished by referencing the sentences via their Gödel numbers. But assuming we have such a setup, it is easy to see that the desired law is not a logical consequence of the T-scheme. If it were, then it would have to be a logical consequence of finitely many instances of the T-scheme, which it obviously is not. This was just the criticism I made of minimalism in Section 2.5. The extra twist here is that, since the law $(\forall A)\mathbb{T}[\langle A \to A\rangle]$ is not derivable from the T-scheme, its negation can consistently be added as an axiom. Thus we can produce a consistent metasystem $PA^\dagger$ which has a truth predicate for PA and in which it is derivable that there exists a sentence A of arithmetic for which $A \to A$ is not true.

There does not actually exist such a sentence, but $PA^\dagger$ does not know this.[11] Its error arises from a nonstandard interpretation of the Gödel numbering system which indexes the sentences of arithmetic.

Here the purely formal nature of Tarski's definition becomes a serious liability. From an external perspective we can easily prove that, in fact, for any arithmetical sentence A, $PA^\dagger$ derives $\mathbb{T}[\langle A \to A\rangle]$. But we cannot affirm $(\forall A)\mathbb{T}[\langle A \to A\rangle]$ from this perspective because the symbol $\mathbb{T}$ *has no meaning* externally to $PA^\dagger$. This is essential to Tarski's characterization of truth predicates, which hinges solely on what is derivable within a given metasystem. And within $PA^\dagger$ the law $(\forall A)\mathbb{T}[\langle A \to A\rangle]$ *fails*.

The only conclusion to be drawn is that the implication law is not a feature of Tarski's characterization of truth. In some metasystems it will be derivable and in others its negation will be derivable. This seems like a serious flaw, given that one of the main reasons we want to have a truth predicate is so that we can make general assertions of precisely this sort. It shows that Tarski's definition of truth "fails to capture the intuitive notion of adequacy he is after".[12]

The analogous difficulty with a Tarskian characterization of falling under has implications for second-order logic. To see the problem, suppose we are setting up a formal system for reasoning abstractly about objects and concepts (or objects and predicates; the concrete versus abstract distinction remains inessential). We will have two types of variables, one for objects and one for concepts. The goal is to draw general conclusions about them. That is what second-order logic is all about.

Now in order to make a concept variable function assertorically we need to use a falling under relation. That is, if x is an object variable and C is a concept variable, we cannot directly assert the juxtaposition of C with x. This expression is not an assertion. Since the variable C refers to, but is not itself, a concept, the way to assert the statement formed by combining

C with x is to say that x falls under C or that $C[[x]]$ is true. This sort of indirect assertion is just the kind of thing that a falling under relation is used for.

But the Tarskian approach to falling under is unsuited to this purpose. First of all, it only licenses us to talk about the concepts appearing in some specific target system, not concepts globally, a point I already made in Section 2.7. In the arithmetical setting, for instance, we could reference arbitrary arithmetical concepts via the Gödel numbers of arithmetical formulas with one free variable. Secondly, even with this restriction we may be unable to affirm even the simplest general statements about concepts and objects. For instance, we would like to affirm an implication law for falling under which says that for any concept C, if x falls under C then x falls under C. A Tarskian characterization of falling under would require of a metasystem that it derive every instance of the scheme

For all x, x falls under $\langle A[z] \rangle$ if and only if $A[x]$

(see Section 2.2). So, since this scheme does not imply the implication law, there will be Tarskian falling under relations for which the implication law derivably fails. As with truth, Tarski gives us no principled grounds for rejecting falling under relations which derivably violate the implication law.

In order to do second-order logic we need to be able to talk about objects falling under concepts. But if all there is to falling under is the derivability of a falling under scheme in some metasystem, then the most basic general logical laws about objects and concepts have no justification.

What about using PA^T in some way as a model for the general notion of a truth predicate? By giving us a basic ability to reason about truth, it does much better than the bare T-scheme. However, much of its power comes from the fact that it gives us the ability to reason inductively about truth, and the problem is that this ability could be strengthened further by importing induction statements on longer, transfinite, well-ordered sets. This explains why, although we can derive in PA^T that anything derivable in PA is true, and even that anything derivable in $PA' \cong PA + \mathrm{Con}(PA)$, is true, we can find a system PA^α higher up in this hierarchy such that the fact that everything derivable in PA^α is true is not derivable in PA^T. The point is that while PA^T matches our intuition notion of truth better than the bare T-scheme does, it can be improved further, and there does not seem to be any principled way to draw a sharp boundary between those predicates that should count as truth predicates and those that should not. In some way the T-scheme alone really should suffice, because it does completely characterize the property of being a truth predicate in the sense that this property is captured by the statement that every instance of the T-scheme is true — or it would be, if there were only some way to say this.

Chapter 4

Assertibility

4.1 Assertibility as a concept

The assertible liar paradox. Formalism. Semantic proofs versus formal derivations. Proving axioms.

The word "proof" has two mathematical senses. It can refer to a syntactically correct derivation in some formal system, or it can have the semantic sense of a conclusive demonstration. I will use the latter meaning exclusively, and reserve the word "derivation" for the syntactic notion. Say that a sentence is *assertible* if it has a proof, in this semantic sense.

Just as with truth, we come to the subject with a pretheoretic notion already in mind, or at least we think we do. It certainly seems as if there are cases where we can know that a statement is true purely by deduction, and that this is somehow qualitatively different from empirical knowledge. But also as with truth, our pretheoretic notion of assertibility is apparently vulnerable to paradox. Consider the *assertible liar sentence*

This sentence is not assertible.

Call this sentence L'. Assuming that L' is assertible seems to lead immediately to a contradiction, and this observation would seem to constitute a proof that it is not assertible. But that would mean that we have proven L' itself, and hence that L' is assertible after all, which is absurd.

One of the main conclusions of Chapter 2 was that the liar paradox shows our pretheoretic global notion of truth to be incoherent. There simply is no such concept. What we have globally is not a concept but a schematic condition, the T-scheme, that we want a truth predicate to implement. The liar paradox then becomes merely a technical obstacle to the construction of truth predicates in limited settings.

The situation with assertibility is a little different because, while the T-scheme is essential to both versions of the paradox, it is not constitutive of assertibility in the way that it is of classical truth. Whether there is a version of the T-scheme for assertibility is a question that we must consider carefully, and this raises the possibility that the assertibility liar paradox might have a resolution that is not available for the classical liar paradox. But before we get to that point, let us start by asking whether our pretheoretic notion of assertibility might, like truth, simply be a mirage. If it is, then there is no problem to solve. This informal discussion will also help to clarify the scope of the analysis which follows.

There are a range of possible views on the significance of assertibility. My aim here is not to be comprehensive but merely to sketch some of the main positions and to indicate where they run into difficulties.

For example, an extreme version of formalism does deny that there is any genuine notion of assertibility in mathematics. This view of mathematics is sometimes encapsulated in the phrase "a formal game with symbols". Certainly, if that is all there is to the subject then the idea of rational demonstration has no place in it. But this brand of formalism is generally discredited. It is telling that the characterization of mathematics as a game played within formal systems tends to be followed by a comment to the effect that the only serious question is whether the axioms one is using are consistent. But since the consistency of a formal system is itself a mathematical question, comments like this unwittingly reveal that the extreme formalist does, after all, accept that at least some mathematical claims can have genuine content. Indeed, a decent variety of number theoretic problems can be straightforwardly posed as problems about consistency. (Add to PA the axiom that every even number greater than two is a sum of two primes. Is this system consistent? Hint: if there is a number that violates the new axiom, then that fact is derivable in PA.)

Hilbert's formalism was subtler than this. His idea was that one could assuage doubts not about all of mathematics, but specifically about those (large) parts of the subject which make possibly controversial assumptions about infinity, by axiomatizing them and then treating the resulting formal systems from a combinatorial point of view.[1] He retained a finitistic core of genuinely meaningful mathematics, so that the question whether a given formal system is consistent could be considered as meaningful at some level. This version of formalism is no obstacle to theorizing about assertibility. All we need is the admission that in some contexts there is a genuine semantic notion of valid proof. Whether infinitary reasoning is legitimate is a separate question, certainly of great interest but not crucial for the theoretical analysis of assertibility as a concept.

Yet another possible version of formalism might consider sufficiently finitistic parts mathematics to be meaningful, but only in some empirical sense that is not amenable to deductive reasoning. Rather than argue this point, I will just propose Euclid's proof that there are infinitely many primes as a good test case. Whether there are infinitely many primes is not a question that can be answered empirically. If one accepts that there are infinitely many primes, and that we know this by deductive reasoning, then I think one has to acknowledge that proof and assertibility are meaningful concepts.

Perhaps assertibility is a meaningful notion but not a precise one? No doubt our intuitive notion of a valid proof is not as transparent as, say, our intuitive notion of a natural number. However, it is essential to the idea of a proof that one is *compelled* to accept its conclusion, and this does not seem to leave any room for imprecision. In any case, lack of precision is a problem to be solved, or at least clarified, by formal analysis, so we can come back to this question later.

A point of view that many mathematicians might find appealing is that assertibility is a genuine concept that is exactly captured by the syntactic characterization of formal derivability. The soundness and completeness theorems (Theorem 3.1 and Corollary 3.5) may be cited here. In the face of these results, what room could there be to deny that formal derivability and semantic provability coincide?

This argument sounds good at first, but in fact it fails badly, for a reason I have already pointed out. The soundness theorem only applies to languages which are interpreted in set models, and restricting ourselves to those settings is not an option, because this would exclude the soundness theorem itself. We consider the soundness theorem to be a *theorem* which we have *proven*, yet it cannot even be formulated in the type of modelled language to which it applies. This is not merely a question of circularity or infinite regress: it is not as though the soundness theorem, once accepted, could be taken to justify the sort of reasoning used in its own proof. Since it makes a statement about all languages that are interpreted in set models, it needs to use a variable that ranges over all sets and hence cannot itself be expressed in a language that is interpreted in a set model. Expanding the soundness assertion to languages that are interpreted in class models is no help, as then we would need a variable that ranges over all classes and this cannot be accomplished in a language that is interpreted in a class model.

In short, the soundness and completeness theorems only give us reason to reduce assertibility to formal derivability in the setting of languages interpreted in set models. But even in this setting, there is another rather interesting difficulty for the identification of semantic proofs with formal derivations. Namely, formal derivability must proceed from a set of axioms and this raises the question of where our knowledge of these axioms comes from. For instance, if the assertible sentences in the language of arithmetic are to be identified with those sentences that are formally derivable, then we need axioms for arithmetic, and evidently our acceptance of these axioms does not come from having formally derived them. (Or if it does, via a derivation in some other formal system, this simply removes us one step to the question of how we know to accept the axioms of that system.)

The problem is especially acute in light of the incompleteness theorems, which seem to show that we have an open-ended ability to find new axioms for arithmetic. It is worth looking at this point a little more closely, because our ability to generate new axioms can be analyzed. Once we know that some axioms for arithmetic (the Peano axioms, say) are true — are evaluated to 1 by the truth function τ, if one prefers — they can be augmented by an arithmetical formulation of the assertion that they are consistent. The justification for this move is an appeal to the soundness theorem. One could avoid invoking the full soundness theorem by making this argument in PA^T, but that still involves the concept of arithmetical truth. PA^T could then be transcended by introducing the concept of truth-in-the-language-of-PA^T. The source of our open-ended ability to find new axioms of arithmetic is thus an open-ended ability to introduce and reason about new concepts, and this is something which is manifestly not captured by the notion of formal derivability. Thus, "the question whether a certain statement is provable cannot be given a mathematically definite formulation since we cannot forsee in advance all possible forms of argument that might be used in mathematics."[2] (But that is different from saying that the concept of provability is imprecise; see Section 4.4.)

I can think of two ways that one could try to overcome the problem of having to prove one's axioms, in defense of the claim that assertibility coincides with formal derivability. The first is to deny that the axioms of arithmetic are, strictly speaking, assertible. On this view, we have a well-defined conception of the set of natural numbers, and the Peano axioms reflect our beliefs about this set, but they are merely beliefs and do not in fact have a perfect rational justification. So what we actually *prove* are statements that the Peano axioms imply this or that conclusion, and we believe with near certainty that those conclusions hold of the natural numbers because we believe with near certainty that the Peano axioms hold of the natural numbers.

This suggestion sounds better in the abstract than it does in specific cases. It would entail that we do not have a perfect rational justification of the claim that $2 + 2 = 4$ but we do have a perfect rational justification of the much more complicated claim that the Peano axioms imply that $2 + 2 = 4$. I return to the test case I proposed earlier, Euclid's proof that there are infinitely many primes. On the view we are considering now, we do not actually know by deductive reasoning that there are infinitely many primes, it only seems that way.

The other possible tactic is to deny that we really do have a definite conception of the natural number sequence. I find this position hard to understand, but it does have its defenders.[3] The idea seems to be that it is the axioms we adopt, and them alone, which specify the structure we are reasoning about. Thus, when one proves a theorem from the Peano axioms one is not proving anything about "the" natural numbers, but only about any model of PA. We can make more precise the class of models we are reasoning about by adding either Con(PA) or its negation as an axiom, but there is no absolute structure of "the" set of natural numbers that determinately does or does not satisfy Con(PA).

Clearly, this view is again a version of formalism. What is strange about it is that it evidently regards the concept of a formal derivation itself as ambiguous. For a formal derivation can have any finite length, so if our concept of a natural number is indefinite then our concept of a possible length of a derivation is also indefinite. Thus, the question of whether Goldbach's conjecture, say, can be settled in PA may not have a definite answer — it could have a derivation whose length is allowed on some versions of "natural number" but not on others. The point is that it is incoherent to claim that our conception of the natural numbers is ambiguous but that our conception of formal derivations is not, because the latter presumes the former.

Indeed, on this view it would seem as though we can establish that the question of whether Con(PA) can be derived in PA does *not* have a definite answer. For if we sharpen our concept of the natural numbers by augmenting PA with Con(PA), then Theorem 3.9 shows that Con(PA) is not derivable in PA. But if we sharpen our concept of the natural numbers by augmenting PA with ¬Con(PA), then the existence of an inconsistency becomes, on the resulting, more precise concept of number, a finite combinatorial fact and therefore one which can be derived in PA. But then why did we need to augment PA if it could already prove ¬Con(PA)? Maybe there is some way of untangling this mess, but it seems to me a long distance to go in order to justify denying the patent fact that we do have a clear, definite conception of the structure of the natural number sequence.

4.2   Proofs and meaning

Uncomputable truth values. Meaning via truth conditions. Meaning via proof conditions. Universal quantification.

We cannot go any further with assertibility without confronting the problem of languages for which we are unable to construct a truth function. When the truth values of its sentences can be mechanically evaluated, a language can be understood in a straightforward computational way. What can we say when this is not the case?

For example, I emphasized in Section 1.5 that the truth function for arithmetic requires infinitely long computations. So from a finitistic standpoint its definition is illegitimate. As far as finitists are concerned, the truth values of general arithmetical sentences cannot be straightforwardly computed, even in principle. This raises questions about what kind of meaning a finitist could ascribe to such sentences. But these sorts of questions are not just a problem for finitists. Predicativists, who accept countably long computations but not uncountably long ones, face essentially the same issue in a different context. They would not have any objection to the definition of the truth function for arithmetic, but they would have a completely analogous difficulty in relation to a language that allowed references to arbitrary real numbers. Here the idea of, say, systematically examining all real numbers to see whether they all had some property would not be seen as legitimate. Even if the property itself were decidable — i.e., we had no trouble checking whether a particular real number had it — there would be no possibility, even in principle, of searching through *all* real numbers on the predicativist view. So what sense can predicativists make of sentences whose truth cannot be evaluated without performing such a search?

For platonists, the same difficulty arises when one considers sentences that quantify over all sets. Surely the idea of performing a computation which checks *all sets* for some property and returns an answer afterwards is not cogent. The fact that the computation had finished would imply that the family of sets which had been examined must itself constitute a set, so there would have to be sets that had not been included in the search. It would be like running through the ordinals, checking them all for some property, and then returning an answer at an additional final step: the final stage would simply represent a new ordinal. So even on the most liberal version of platonism, it is in principle impossible to mechanically calculate the truth values of sentences which quantify over all ordinals. Thus, it seems that on any coherent foundational stance one is faced at some point with making sense of sentences whose truth cannot be mechanically checked (under that stance's interpretation of "mechanically checked").

Nonetheless, it seems clear that at least some sentences of this type can be understood and even proven. A finitist should agree that every natural number is either even or odd; a platonist should agree that every set is a subset of itself. The question is "in what the knowledge of the condition under which a sentence is true can consist, when that condition is not one which is always capable of being recognised as obtaining".[4]

Intuitionists answer this question in terms of assertibility.[5] This is the key difference between them and ordinary finitists, but it is completely independent from their rejection of infinite computations. The conception of proofs as the source of meaning neither implies nor is implied by views about the cogency of infinite computations. It could apply equally well to the predicative or set-theoretic settings.

The meaning-via-proofs view is sometimes carelessly expressed by saying something to the effect that when one asserts a sentence one is *really* asserting that that sentence can be proven.[6] This is not a good formulation because if it is meant to be universal then it creates an obvious infinite regress. Moreover, its application to complex sentences is ambiguous: is a sentence of the form $A \wedge B$ to be understood as really asserting that $A \wedge B$ can be proven, or that it can be proven that A can be proven and that B can be proven? What if A and B themselves have internal structure?

Erasing the distinction between "37 is prime" and "'37 is prime' can be proven" is a recipe for confusion. One is a statement about a number, the other a statement about a sentence. A more careful version of the intuitionist's dictum is that "the *sense* of a mathematical assertion denoted by a linguistic object A is intuitionistically determined (or understood) if we have laid down what constructions constitute a *proof* of A".[7] Thus, to understand the sense of the sentence "37 is prime" is to know how to recognize a proof that 37 is prime, but we do not regard "37 is prime" as actually being the sentence "it is provable that 37 is prime" in disguise.[8]

The analogous error in the classical setting is the idea that when one asserts a sentence one is *really* asserting that that sentence is true.[9] Perhaps this is a less serious mistake, because once a truth predicate is available the sentences A and $\ulcorner$ "A" is true$\urcorner$ are equivalent by definition. But it is connected to the error we discussed in Section 1.7 of thinking that truth is somehow a component of meaning. Recall that the source of this idea was the principle that the meaning of a sentence is given in the conditions under which it is true (and the error comes from failing to realize that truth plays no role here other than to express a schematic claim). We uncritically granted this principle in Section 1.7, but let us now examine it more closely. In fact it is perfectly tautological. Is the meaning of the sentence "snow is white" given in the conditions under which "snow is white" is true, or more simply, in the conditions under which snow is white? Yes, of course it is, because the condition under which snow is white is simply that snow should be white. A similarly vacuous statement could be made about any sentence. We can certainly agree that the meaning of any sentence is given in the conditions it expresses, because without further qualification this is just a roundabout way of saying nothing.

Perhaps the intention behind identifying meaning with truth conditions is that we want the truth conditions for a new sentence to be given in terms of sentences that are already understood. If so, what we are doing is endowing sentences with meaning by showing how to translate them into previously understood sentences. This is perfectly legitimate, although there is little need to describe the process in terms of "truth conditions": all we are talking about now is some version of extending a language by adding defined terms. (We can say that $\ulcorner A \leftrightarrow B \urcorner$ is true provided $\ulcorner A \to B \urcorner$ and $\ulcorner B \to A \urcorner$ are both true, but it would be more direct to just say that $\ulcorner A \leftrightarrow B \urcorner$ abbreviates $\ulcorner (A \to B) \land (B \to A) \urcorner$.) In any case, running the process in reverse, we will eventually reach a minimal level of language which cannot be reduced to anything more primitive. At that point the only way to go further in explaining meaning in terms of truth conditions is by having those conditions be nonlinguistic. In other words, on this view grasping meaning ultimately comes down to a nonverbal ability to compute truth values.

This conclusion brings out the reason why meaning is more problematic for languages which do not have a computable truth function. In these cases there is no straightforward possibility of reducing meaning to some nonlinguistic skill. "The difficulty arises because natural language is full of sentences which are not effectively decidable"[10] (and so is mathematical language).

At this point the intuitionistic approach to meaning might begin to be attractive. Consider the assertion that every set is a subset of itself. Not only are we practically unable to explicitly verify this statement separately for every set there is, we do not even believe that such a verification is possible in principle. The way we know this statement is true is not because we have directly verified it, or even think we could verify it in any sense, but because we know how to prove it, which suggests that the actual content of the statement has to do with its provability. Thus we are led to associate meaning with proof.

The first objection one might have to this proposal is that understanding the concept of provability requires one to have already grasped the concept of truth: "a proof is an argument which shows the truth of a proposition; so if propositions cannot be true independently of being proved, one wonders what a proof has to prove".[11] Now, the appearance of truth in this complaint may be misleading; yes, for any sentence A, a proof of A is an argument which shows that A is true, but we can also express this principle schematically as

A proof of "$\mathcal{A}$" is an argument which shows that $\mathcal{A}$.

For instance, a proof of the sentence "37 is prime" is an argument which shows that 37 is prime, and truth plays no role in that formulation. This is just Dummett's fallacy again, with truth being employed for no purpose other than to express a schematic claim.[12] But there is still an issue here. How do we know what it means to prove that 37 is prime if we do not already understand what "37 is prime" means?

The intuitionist's answer is that we must take the proof relation as fundamental. We know nothing about the meaning of the sentence "37 is prime" until we have said what would count as a proof of it. And once we have said that, we have said everything there is to say about what it means.[13]

However, like the meanings-as-truth-conditions idea, this proposal requires a nonlinguistic basis. If the ultimate source of linguistic meaning is the characterization of what counts as a proof, then this characterization ultimately has to be nonlinguistic, lest we enter an infinite regress. Is it possible to reduce the problem of recognizing valid proofs to the level of a nonverbal skill?

There is no problem mechanically verifying the correctness of a formal derivation, but the open-ended nature of the concept of a valid proof makes the prospects for mechanical recognition seem unlikely. The intuitionistic proposal is to specify what counts as a proof of a sentence recursively on the complexity of the sentence. In the atomic case, what would count as a valid proof would probably vary depending on the subject matter, so let us grant that we are able to recognize these proofs somehow. Then, proceeding recursively, we could take a proof of $A \wedge B$ to just be a proof of A together with a proof of B. If we knew how to recognize proofs of A and B individually, then we could learn to recognize proofs of $A \wedge B$ using this criterion. Gaining this ability need not require any special linguistic skill. A proof of $A \vee B$ could be something with is either a proof of A or a proof of B, and so on.

The fatal problem appears only when we get to universal quantifiers. What would count as a proof of $(\forall x)A[x]$? In orthodox intuitionism, the answer is: a construction which, for every x, generates a proof of $A[\hat{x}]$. But how could such a construction be recognized? For instance, suppose we were given a construction which, for every set X, produces a proof that X is a subset of itself. How would we recognize that the construction actually did this?

Certainly we cannot check the result of the construction separately on every set X. We need to have some way of globally recognizing that it succeeds. Since we cannot do this by mechanically checking each instance separately, and we are assuming we do not have any native grasp of what it means to quantify over a proper class — that is what we are now trying to provide — the only alternative is to do it by having a proof that the construction succeeds for each X. But then how would we recognize that proof? Since the assertion in question is itself universally quantified, according to orthodox intuitionism what we would need is a second construction that, for every set X, produces a proof that the first construction produces a proof that X is a subset of itself. Thus we enter an infinite regress.[14]

This is the point at which the truth-condition account of meaning fell down, when we had to identify the truth conditions for sentences of the form $(\forall x)A[x]$ where x ranges over a proper class. The problem was that any expression of those conditions has to itself involve quantification over a proper class. But the proof-theoretic account of meaning does no better: any characterization of the proof conditions for such a sentence also has to involve quantification over a proper class. So the intuitionist's account of meaning in terms of proof fails just as badly.[15]

It seems that there is something wrong with the whole idea of formulating a theory of meaning in the sense of this section; the meaning of a sentence, or at any rate of a universally quantified sentence, in general cannot be expressed in any more elementary terms than it already is in that sentence. The meaning of "every set is a subset of itself" cannot be explained in terms of the obtaining of some truth conditions, because those conditions cannot be expressed any more simply than they are in the original sentence: the condition under which "every set is a subset of itself" is true is just that every set should be a subset of itself, and there is no more elementary way to say this. Nor can it be explained in terms of the existence of a proof of the sentence, because it is, if anything, harder to understand what would constitute such a proof than it is to understand the sentence itself.

4.3 Existence

Classical and constructive views on existence. Proofs which use the axiom of choice. Proofs which use the law of excluded middle. Quantifying over proper classes. Asymmetry between a sentence and its negation. Ontology.

There are two fundamental questions which distinguish the main schools in the philosophy of mathematics. The first is to what extent infinite objects or constructions are legitimate. How this question is answered distinguishes finitism (all objects and constructions are finite), predicativism (all objects and constructions are countable), and platonism (objects and constructions can be uncountable). The second is whether an object having some property could exist without this fact being, in principle, discoverable, which distinguishes the classical and constructive points of view. These two axes are independent.

Both the classical and the constructive theories of meaning discussed in the last section bear on the second question. On a classical theory of meaning, the content of an existence assertion is to be found in some truth conditions which presumably make no reference to what could or could not be known by any observer. So whether it is discoverable that some object has some property would not be part of what it means for some object to in fact have that property. On a constructive theory of meaning, in contrast, the content of an existence assertion is to be found in a characterization of what would count as a proof of the assertion. This makes the idea of such a statement being true yet impossible to prove simply nonsensical. That is, the constructive approach to meaning makes knowability essential to truth.

However, as we saw in the last section, neither of these theories of meaning works. The content of an existence assertion does not reside in either truth conditions or proof conditions: it resides in the sentence itself. Thus, theories of meaning cannot help us choose between the opposing conceptions of the nature of reality and language on which the classical and constructive philosophies are based.

Both conceptions have some intuitive plausibility. The classical view is that there is a well-defined, objective, external reality, that language relates to that reality in a straightforward way, and that rational agents play no essential role in this picture. While to the constructivist, the idea of an unknowable fact is a metaphysical fiction, like an invisible color: assertions that in principle cannot be verified simply have no content. Let us acknowledge the appeal of both positions while recognizing that the theories of meaning that were supposed to compel us to accept one of them over the other do not succeed in doing so.

There is actually less practical difference between the two views than one might think; in most ordinary settings it is not even clear that they disagree in any substantive way. This may sound like a surprising claim, since one is used to hearing about the dramatic differences between classical and intuitionistic mathematics. However, these differences have much more to do with the finitistic aspect of intuitionism than with its constructive conception of mathematical existence. Let me explain this point by discussing two kinds of examples.

One class of examples involves proofs using the axiom of choice, the principle that given any family of nonempty sets, it is possible to select a single element out of each set in the family. Such proofs are generally considered to be "nonconstructive", but "nonexplicit" might be a better description. Indeed, it seems to me that one of the main reasons the axiom of choice is so believable is because one imagines that in principle one actually could go through the entire family of sets, selecting one element from each. And a principal source of skepticism toward the axiom is probably doubt about the legitimacy of the idea of making infinitely many arbitrary choices. The point is that this is not primarily a problem about whether there are objects which cannot be constructed; it is a problem about what counts as a valid construction.

Another kind of allegedly nonconstructive existence proof is exemplified by the famous proof of Jarden that there exist irrational numbers a and b such that a^b is rational.[16] Consider the number $\sqrt{2}^{\sqrt{2}}$. If it is rational then we are done. Otherwise $\left(\sqrt{2}^{\sqrt{2}}\right)^{\sqrt{2}}$ has the form a^b with a and b irrational, and $\left(\sqrt{2}^{\sqrt{2}}\right)^{\sqrt{2}} = \sqrt{2}^2 = 2$, so again we are done. This argument shows that either $\sqrt{2}^{\sqrt{2}}$ or $\left(\sqrt{2}^{\sqrt{2}}\right)^{\sqrt{2}}$ is an example of the desired phenomenon, but it does not tell us which one. (In fact, it is known that $\sqrt{2}^{\sqrt{2}}$ is irrational, but that requires a different and much harder argument.)

This is indeed an example of an intuitionistically invalid argument, but it trades heavily on the intuitionist's rejection of infinite computations. If infinite computations are allowed, then we can easily determine which of $\sqrt{2}^{\sqrt{2}}$ and $\left(\sqrt{2}^{\sqrt{2}}\right)^{\sqrt{2}}$ is irrational. How does this relate to the proof given above? It does so through the appeal to the law of excluded middle: either $\sqrt{2}^{\sqrt{2}}$ is rational, or it is irrational. If pressed to justify this dichotomy, one might say: just compute the decimal expansion and look to see whether it is eventually periodic. And this would be a perfectly valid constructive justification but for the reference to an infinite computation. That is, an intuitionist who could be persuaded to accept infinite computations would then accept the use of excluded middle in Jarden's proof, and would therefore accept the proof. It would be constructively valid because it would, in principle, tell us how to find an example of the desired type: first determine whether $\sqrt{2}^{\sqrt{2}}$ is rational or irrational, then follow the relevant branch of the argument.

This illustrates the point that the restrictiveness of intuitionistic mathematics as compared to classical mathematics has more to do with the nonacceptance of infinite computations than with views about the nature of mathematical existence. Provided we accept infinite computations, interpreting existence constructively would seem to have little effect on ordinary mathematics, because nearly all the assertions of interest in ordinary mathematics have decidable truth values under a sufficiently liberal interpretation of "decidable". This is not a new point but it deserves to be better known. Whereas if infinite computations are rejected, then the relevant comparison is not between intuitionism and classical mathematics but between intuitionism and finitism.

The preceding discussion also brings out the relation between constructive views on existence and the law of excluded middle. Let A be a sentence. If we know that A has a definite truth value, either 0 or 1 — even if we do not know which — this gives us clear grounds for affirming the sentence $A \vee \neg A$; and if we doubt that A has a definite truth value, then we should equally doubt the assertion $A \lor \neg A$. So what underlies the law of excluded middle is an existence question: do unknowable truth values exist?

This observation can help us identify situations where classical and constructive views of existence conflict, even when infinite computations are accepted. For example, suppose that we know how to determine, for each ordinal, whether it has some property P. Are we then licensed to affirm that either every ordinal has P or some ordinal does not? The point is that checking all ordinals for some property is, for the platonist, like checking all numbers for some property is for the finitist. We already considered this kind of situation in the last section, when we discussed settings in which a truth function cannot be mechanically evaluated because of quantification over a proper class. The issue before us now is whether we should accept that sentences like "every ordinal has the property P" have definite truth values, even if there might be no conceivable circumstance in which we could know what they are, not even with the help of infinite computations.[17]

The classical impulse is to say that we should accept this, on the grounds that there is a fact of the matter: either every ordinal has the property, or some ordinal does not. The intuitionistic impulse is to say that we should not, on the grounds that the idea of a fact which in principle cannot be known is not cogent. Part of the insuitionist's motivation could also come from unease with the open-ended nature of the ordinals: no matter how many of them one has examined, there will always be more to check, lending the concept of "all ordinals" an indefinite quality. Again, this disagreement can be seen as an aspect of the fundamental dispute over the nature of existence, where in this case the truth value of a sentence is the object whose existence is in doubt.

The intuitionistic position tends to be seen as a rejection of the idea of objectivity in mathematics, but it does not have to be understood in this way. It is not as if we have any freedom to choose the truth values of mathematical assertions. An analogy with the universe of sets may help here. The set-theoretic universe is open-ended in the sense that there is no conceivable circumstance in which every set is available: one always has the capacity to generate new sets. We can think of truth values in the same way. We have the capacity to generate them, either via deductive reasoning or by straightforward mechanical calculation, but there is no maximal state in which all truth values are available.[18] This does not mean that we have any freedom to decide which truth value any particular sentence should have, any more than we have any freedom to decide which sets exist: "it is consistent with anti-realism in this sense to retain the idea that a discourse is representational, and answers to states of affairs which, on at least some proper understandings of the term, are independent of us".[19]

Let me also note that the searching-through-the-ordinals example reveals an asymmetry in the logical operation of negation. Normally we think of A and $\neg A$ as standing on an equal footing, so that the classical law of double negation elimination, $\neg\neg A \to A$, is automatic. But consider a sentence of the form $(\exists X)A[X]$, where A represents a decidable property and X ranges over all sets. A sentence like this has some chance of being verified by a mechanical search, but no chance of being falsified in that way. In other words, the existence of a set with some decidable property can be seen to be true by inspection, whereas the nonexistence of such a set could only be deduced indirectly, never directly observed. In this way the two cases are not symmetric. Our intuition for there being a symmetry between any sentence and its negation in a consequence of our intuition that truth values always exist: if every sentence has a truth value, then negations just have the opposite truth values and the relation between any sentence and its negation is symmetric. But if truth values are not computable and there is a question of whether they can meaningfully be said to exist, then this also calls into question the presumed symmetry between A and $\neg A$.

There are also cases where neither a sentence nor its negation could ever be mechanically verified. For instance, consider an assertion of the type: for every set X there exists a set Y to which it bears some relation R. Even if we could determine whether the relation holds of any given X and Y, no set of observations could assure us that every X has the stated property — we could not have inspected every possible value of X — nor could any set of observations assure us that any particular set X failed to have it — we could not have inspected every possible value of Y. The truth value of such an assertion might be thought of as the fictional result of a computation that never actually terminates, a formulation that perhaps sharpens both the intuition behind thinking that this truth value legitimately exists and the intuition behind doubting this.

I feel that the ontological aspect of the law of excluded middle has not been sufficiently appreciated. As we saw above, whether we accept the law of excluded middle hinges on our attitude toward the existence of truth values which in principle cannot be determined. So it can be understood as a question of ontology, of the existence of something which we might be unable to locate. Now, postulating the existence of truth values might seem inconsequential since the only possible truth values are 0 and 1. But this attitude is mistaken: to the contrary, the existence of unknowable truth values is an ontologically rich assumption. For example, let R be some decidable relation between numbers and sets and consider the formula $A[n] =$ "$(\exists X)R[[n, X]]$" which asserts of any given natural number n that there exists a set X to which it bears the relation R. For each number n, the truth value of $A[\hat{n}]$, if it exists, is either 0 or 1. So to affirm that all these truth values exist is to affirm that a particular infinite sequence of 0's and 1's exists; that is, it gives us a subset Y of $\mathbb{N}$. This shows that assuming truth values exist yields *a substantive new set existence principle*, a new means of generating sets which might not be generated in any other way. What may be unnerving about the set Y in this example is that even using uncountably long computations, we might have, in principle, no way to determine whether a given number belongs to it.

Thus, after controlling for varying positions on the acceptability of infinite computations, we find that the most significant difference between classical and constructive views on existence lies not in their consequences for ordinary mathematics, which are minimal, but in their consequences for the existence of exotic sets of the preceding type.

4.4 *A* versus $\mathbb{A}(A)$

Paradoxes of the proof relation. Validity of capture. Self-affirming nature of release. Informal characterization of proofs. The logical tightrope. Comparison between $\mathbb{A}$ and $\mathbb{D}_S$.

In the last section, we asked whether we have a right to assume that sentences which involve quantification over a proper class necessarily have well-defined truth values. Assuming that they do leads to set existence principles that would not otherwise be available, but unless those new principles lead to a contradiction this in itself does not compel us to adopt one view over the other.

Something like this actually does happen in other settings. As we will see, assuming that the law of excluded middle holds for sentences involving the concept of assertibility leads to absurd consequences. This might not be so surprising, since the open-ended nature of assertibility, which we briefly touched on in Section 4.1, already suggests that the question of whether a given sentence is assertible might not have a well-defined truth value. The assertible liar paradox (p. 89) points toward the same conclusion: it is hard to believe that the sentence L' has a definite truth value.

Many authors have felt that it may be appropriate to use different forms of logic in different settings.[20] Thus one could easily take the position that classical logic is appropriate in mathematics proper, while intuitionistic logic is appropriate for reasoning about assertibility. But is the source of potential indeterminacy the same in both cases? The assertible liar sentence L' says of itself that it cannot be proven, i.e., there does not exist a proof. In this existence statement we are quantifying over an infinite set of proofs, and if proofs can be infinitely long, maybe even over a proper class of proofs. So it looks as though uncomputable quantification could be the culprit here, just as before. However, we can formulate similar paradoxes that involve only quantification over finite sets, or even no quantification at all. For example, let A be the sentence

  p is not a proof of A

and let p be the sentence

  If p were a proof of A, then it would be a proof that p is not a proof of A, a contradiction, so p cannot be a proof of A.

No quantification appears in either p or A. So is p a proof of A? It certainly appears to be a valid chain of reasoning whose conclusion is precisely what is expressed by A. But this is absurd.

I repeat the comment that the use of explicit names, here p and A, could be avoided by a syntactic construction of Quinean type. Or we could finesse the issue by reformulating A as the "Fermat paradox"

  There is no proof of this sentence that is short enough to fit in the margin.

If every string of characters short enough to fit in the margin definitely is or is not a proof of the displayed sentence, then either one of them is a proof or none of them is. The quantification is finite, so even intuitionists should accept this inference. But the first alternative leads to a contradiction, and therefore the second must be the case. Thus, we have proven the displayed sentence, and the proof was short enough to fit in the margin. Contradiction.

(The vagueness of the phrase "short enough to fit in the margin" is not important. It could be replaced by an exact description such as "at most 1,000 characters long" without affecting the paradox.)

The obvious conclusion to draw from examples like these is that we cannot assume that well-defined truth values exist when we are dealing with open-ended concepts like assertibility. Thus, let us adopt the "safe" constructivist position that unknowable truth values do not exist.

This is mainly a negative position in that it prevents us from affirming all instances of the law of excluded middle, as well as any other principle whose validity hinges on reasoning involving objects whose existence might never be known. It is therefore "safe" because its main effect is to block potentitally illegitimate reasoning. However, in one respect it is a positive position. Letting $\mathbb{A}$ symbolize the clause "it is assertible that", we can affirm the implication

$$A \to \mathbb{A}(A)$$

for any sentence A: if snow is white, then it is assertible that snow is white, and so on. This is one law that constructivists can affirm but classical mathematicians cannot.[21] Indeed, this law effectively characterizes the constructivist position. It does not seem very dangerous because it appears to be incapable of being contradicted — we do not expect to ever find ourselves in the position of having a sentence A for which we are able to prove both A and $\neg\mathbb{A}(A)$ (i.e., both A and "A is not assertible").[22] (But see Section 4.5 for a qualification of this statement.)

The implication $A \to \mathbb{A}(A)$ is sometimes called the *capture* law. What about *release*, the reverse implication $\mathbb{A}(A) \to A$?

In orthodox intuitionism both capture and release are accepted.[23] Because intuitionists regard proof as constitutive of meaning, on their view there appears to be a straightforward equivalence between a sentence A and the assertion that there is a proof of A. But this is not perfectly clear because it seems to assume a constructive interpretation of A and a classical interpretation of "there is a proof of A". On one side of the alleged equivalence we insist that all of the content of A resides in a specification of what counts as a proof of A, but then on the other side we take the statement "there is a proof of A" at face value. If it is also interpreted constructively then the equivalence becomes problematic.

To see the difficulty, let us ask how an intuitionist would prove the release law. According to the standard intuitionistic account of implication, a proof of this law would manifest as a procedure for converting any proof of the premise (namely, there exists a proof of A) into a proof of the conclusion (namely, A), and a proof of the premise would manifest as a construction of an object p together with a proof that p proves A. The obvious procedure to convert this into a proof of the conclusion, i.e., a proof of A, would be to discard the second part and return the construction of p as output.

But does this actually convert any proof of the premise into a proof of the conclusion? Only if we know that p actually is a proof of A. Thus, in order to be sure that this construction works we need to be able to infer "p proves A" from "there is a proof that p proves A" — which is just a special case of the implication we are trying to prove, with "p proves A" in place of A.

As we saw in Section 4.2, there are other circularity issues in the proof theory of meaning, which is why we rejected that theory. However, the preceding analysis is still helpful because it brings out the self-affirming nature of the release law. In effect the law affirms that all proofs are sound. This is circular, and as the assertibility liar paradox shows, viciously circular, because of the possibility that the release law itself might be employed in a proof whose soundness it affirms. It is analogous to the sentence $G^*[\hat{g}^*]$ discussed in Section 3.5 which affirmed the consistency of the system PA^* obtained by adding $G^*[\hat{g}^*]$ itself to PA as an additional axiom. As we saw there, we have a perfect right to add a new axiom which affirms the consistency of the previously accepted axioms, but a sentence which affirms the consistency of the previous axioms together with itself is a recipe for inconsistency.

Actually, the particular sentence $G^*[\hat{g}^*]$ is more analogous to the special instance $\mathbb{A}(\bot) \to \bot$ of the release law, which can also be written as $\neg\mathbb{A}(\bot)$. This single instance — call it the *law of assertible noncontradiction* — is already problematic. Of course we do not want there to be a proof of $\bot$, but there is something suspicious about quantifying over all proofs in affirming that there is no proof of $\bot$ and then allowing this affirmation to be used as an ingredient when building proofs. After all, the whole point of affirming the law of assertible noncontradiction is to make it available to be used in proofs.

Before we continue this discussion it will help to get a clearer idea of what proofs are. Now, because of the open-ended nature of assertibility, we cannot expect to find a really explicit characterization. Presumably, recognizing the correctness of any explicit characterization would then enable us to recognize as valid new proofs not covered by it.

Still, informal characterizations can be informative. Adapting an idea of Wright,[24] we could say that A is assertible if any sufficiently intelligent rational agent would, given sufficient time, accept that A can be conclusively demonstrated. As vague as this characterization is, it is still helpful. For instance, it suggests that we were premature in concluding, earlier in this section, that bare assertions of proof do not involve quantification over proper classes. "Sufficiently intelligent" and "sufficient time" could be understood by a finitist as allowing arbitrarily large finite resources or by a platonist as allowing resources of any cardinality, but in either case involving a quantification which is uncomputable from the given perspective. Thus, issues about the existence of truth values for sentences involving assertibility might, after all, be understood as arising from uncomputable quantifiers.

Earlier I raised the question of whether the notion of assertibility is a precise concept. Certainly the preceding formulation is not precise, but one could take the position that it is merely an imprecise formulation of a precise concept. A weaker position would be that the concept itself may be imprecise, but it is precise enough to reason about formally. Either way, objectivity is built into the concept of assertibility: if *all* agents with sufficient intelligence and resources agree that something is assertible, then its status in this regard is evidently an objective fact. Again, as I explained in Section 4.3, this does not force us to accept the law of excluded middle for assertibility claims provided we reject the idea of unknowable truth values.

Returning to the law of assertible noncontradiction, we have to ask whether an arbitrarily powerful intellect could ever come to accept that $\perp$ had been given a compelling rational demonstration. It certainly seems unlikely, but can we be sure? One wants to say that there can be no proof of $\perp$ because anything presented as such would automatically be rejected. But proofs tend to be articulated objects and they cannot simply be rejected as a whole without identifying an invalid step. For instance, the directive "reject as invalid any proof of $\perp$" would not tell us what to do if the proof arose, say, by combining a proof that there exists a proper class of measurable cardinals with a proof that there does not exist a proper class of measurable cardinals. One alternative would have to be rejected, but which one? Remember that we can reason about sentences which have uncomputable truth values, uncomputable even in principle. So there might be no canonical way to make this decision.

Certainly, if we grant that every sentence has a well-defined truth value then we can specify the faulty step in any purported proof of $\perp$: it is the first step which derives a conclusion B from a conjunction of premises A, such that the universal closure of the implication $A \to B$ is false. But if we do not grant that truth values exist then we cannot make this argument.

Have we any plausible candidate for a valid proof of a contradiction? Yes, of course: the assertible liar paradox. The flaw in the paradoxical reasoning is not obvious, and I want to suggest that it should be taken seriously as revealing a real possibility that assertibility is a meaningful concept that cannot be reasoned about consistently. This is no more than a possibility: further reflection may convince us that assertibility is not meaningful, or that some step in the derivation of the contradiction is invalid. But we must also acknowledge that further reflection could lead us to the opposite conclusions, that assertibility is a legitimate concept and that every step in the derivation of the assertible liar paradox is valid.

Indeed, there is a school of thought which holds that there are genuine proofs of contradictions.[25] I do not go that far. I merely refuse to rule out the possibility that this position is wrong, which means that I do not take the law of assertible noncontradiction, or the release law in general, as an established fact. *And if the release law is unavailable then the assertible liar paradox cannot be derived.* That is, refusing to deny the possibility that the paradoxical derivation could be valid is precisely what is needed to block the paradoxical derivation. This is the logical tightrope we have to walk when reasoning about assertibility.

Since the preceding point is so crucial, I will repeat it. The implication $\mathbb{A}(A) \to A$ is needed to derive the assertible liar paradox. By not taking this implication as proven — not taking it as given that assertibility is sound — we lose the ability to use the assertible liar paradox to derive a contradiction. The moment we decide that assertibility is sound, we will have to accept paradoxical reasoning as valid, which will then show that paradoxical reasoning is not sound.

I owe a demonstration that assertibility reasoning actually is consistent when the release law is not accepted (and intuitionistic logic is used). That will come in Chapter 5; first we need to axiomatize assertibility, which will be the topic of the next section.

I close this section with a comparison between the predicates $\mathbb{A}$ and $\mathbb{D}_S$. While obviously related, they are conceptually very different. Since $\mathbb{A}$ encompasses all possible means of rational argument, and we adopt the principle that all truths can be known, we are able to affirm $A \to \mathbb{A}(A)$ for arbitrary A. In contrast, because any particular formal system only offers limited means of reasoning, we cannot affirm $A \to \mathbb{D}_S(A)$ except in very special circumstances. But it is just this limitation which gives us the ability to affirm the soundness of reasoning within particular formal systems (as we did in Section 3.6 for Peano arithmetic, for instance). Thus $\mathbb{D}_S(A) \to A$ can be affirmed for various systems S, but a general affirmation of $\mathbb{A}(A) \to A$ is problematic because we would need to prove the soundness of all possible reasoning, not just the reasoning available in some particular formal system.

4.5   Axiomatizing assertibility

Proofs not reducible to constructions. Assertibility as the appropriate
target for axiomatization. Axioms for assertibility. Reasoning under
A. *Interaction with the logical symbols.*

In informal settings — when arguing for new axioms, for instance —
the question of whether a particular block of text represents a rigorous
proof may be fundamentally imprecise. One feels that it could depend
on background knowledge or interpretation. Whether a statement can be
conclusively demonstrated seems a more definite question than whether a
particular demonstration of that statement is conclusive. Thus I will not
attempt to formally analyze the proof relation ("p is a proof of A").

Previous efforts to formalize the proof relation have shown the task to be
complicated and contradiction-prone.[26] Although the intuitionists have an
informal account of proof in relation to the logical symbols,[27] it involves
a further primitive concept, the concept of a "construction", which has
proven difficult to formalize.

The appeal to constructions also involves a fundamental equivocation
between having a proof of A and having a proof that A is provable. As I
have emphasized, it is essential to keep these two ideas distinct. To see the
problem here, consider universal quantification. In intuitionism the way we
prove $(\forall x)A[x]$ is supposedly by recognizing that we have a construction
which produces, for each x, a proof of $A[\hat{x}]$. But suppose x ranges over a
set with a single element, say $\{0\}$; then $(\forall x)A[x]$ should be equivalent to
$A[\hat{0}]$, yet on the intuitionistic account a proof of $(\forall x)A[x]$ would merely be a
construction that we can prove generates a proof of $A[\hat{0}]$. This would make
$(\forall x)A[x]$ equivalent to $\mathbb{A}(A[\hat{0}])$, not to $A[\hat{0}]$. In general, the intuitionistic
idea of a proof of $(\forall x)A[x]$ as a construction which can be seen to generate a
proof of $A[\hat{x}]$ for every x would actually function as a proof not of $(\forall x)A[x]$
but of $(\forall x)\mathbb{A}(A[x])$.[28]

A similar issue arises with implication, where the sentences A and $\top \to$
A ought to be equivalent, but on the intuitionistic account of implication a
proof of the latter would effectively be not a proof of A but a construction
which we can prove generates a proof of A. In general, the intuitionistic
idea of a proof of $A \to B$ as a construction which can be seen to convert
any proof of A into a proof of B would actually function as a proof not of
$A \to B$ but of $\mathbb{A}(A) \to \mathbb{A}(B)$.

If the idea of reducing the concept of a proof to something more basic
— the concept of a construction — fails, then the best explanation of what

a proof is could be nothing more than the trivial scheme mentioned in Section 4.2: a proof of "$\mathcal{A}$" is an argument which shows that $\mathcal{A}$. Once we understand what the logical symbols mean, it appears that there is little else we can say at the axiomatic level about what would count as a proof of a sentence with a given logical structure.

When no further reduction of an informally understood concept is possible, the way to proceed formally is to take it as a primitive notion and try to axiomatize its behavior. I claim that the right notion to axiomatize is not the proof relation (p proves A) but the concept of assertibility (A can be proven).

Assertibility can be given a simple axiomatic formulation.[29] One technical point to be considered first is whether we should regard $\mathbb{A}$ as a logical operator or as a predicate. In ordinary language, the distinction would be between "it is assertible that snow is white" and "the sentence 'snow is white' is assertible". Only the latter interpretation allows us to quantify over the A in $\mathbb{A}(A)$, so it is the one we will mainly be interested in. The logical operator version is slightly easier to state, however. Here we would assume a language in which $\mathbb{A}(A)$ is a formula whenever A is a formula, and we would axiomatize $\mathbb{A}$ using three schemes:

$$
\begin{array}{ll}
(\mathbf{C}_{lo}) & A \to \mathbb{A}(A) \\
(\wedge_{lo}) & \mathbb{A}(A) \wedge \mathbb{A}(B) \to \mathbb{A}(A \wedge B) \\
(\to_{lo}) & \mathbb{A}(A) \wedge \mathbb{A}(A \to B) \to \mathbb{A}(B)
\end{array}
$$

with A and B ranging over all formulas. $(\mathbf{C}_{lo})$ is a capture scheme and $(\wedge_{lo})$ and $(\to_{lo})$ are self-explanatory. If we can prove A and we can prove B, then we can prove $A \wedge B$; if we can prove A and we can prove $A \to B$, then we can prove B. We take the logic to be intuitionistic, for reasons discussed in the last section.

But we will be concerned more with systems which take $\mathbb{A}$ to be a predicate. In arithmetical settings, for example, we will take $\mathbb{A}$, like $\mathbb{D}$ and $\mathbb{T}$, to be a concept symbol that applies to Gödel numbers of sentences, so here the appropriate notation is $\mathbb{A}[\langle A \rangle]$, not $\mathbb{A}(A)$. In order to keep the discussion general we need not specify a particular language at this point. Let us just assume that we are working with some interpreted language, in the sense of Section 2.3, that includes an assertibility predicate and has the resources to quantify over the sentences whose assertibility is discussed.

(If this kind of quantification is not available then the axioms listed below would have to be treated as axiom schemes.) A key point is that the assertibility predicate could be self-applicative: it might apply to sentences in which it appears.

Since the predicate interpretation of $\mathbb{A}$ only applies to sentences, the following axioms are formulated in terms of universal closures. We employ a single scheme, (C) ("capture"), and three individual axioms, $(\wedge)$, $(\rightarrow)$, and (I) ("internal"):

$$
\begin{array}{ll}
\textbf{(C)} & A \rightarrow \mathbb{A}[\langle A \rangle] \quad \text{for any sentence } A \\
(\wedge) & \mathbb{A}[\langle \overline{A} \rangle] \wedge \mathbb{A}[\langle \overline{B} \rangle] \rightarrow \mathbb{A}[\langle \overline{A \wedge B} \rangle] \\
(\rightarrow) & \mathbb{A}[\langle \overline{A} \rangle] \wedge \mathbb{A}[\langle \overline{A \rightarrow B} \rangle] \rightarrow \mathbb{A}[\langle \overline{B} \rangle] \\
\textbf{(I)} & \mathrm{Ax}[\langle A \rangle] \rightarrow \mathbb{A}[\langle A \rangle]
\end{array}
$$

Recall that $\overline{A}$ can be any universal closure of A.

In $(\wedge)$ and $(\rightarrow)$ the quantification is over all formulas A and B. In (I) it is over all sentences A, and the predicate Ax is defined to hold of any sentence which is either one of the Hilbert axioms (listed at the end of Section 3.2) or one of the other axioms of the system, not including (I) itself.

Although we have not included the release law $\mathbb{A}[\langle A \rangle] \rightarrow A$, it might still be adopted in the form of a deduction rule which infers A from $\mathbb{A}[\langle A \rangle]$. This rule could be justified by starting with a formal system S that does not include the release rule, then proving a soundness principle which states that whenever $\mathbb{A}[\langle A \rangle]$ is derivable in S, A is assertible. (This might be done using a Tarskian truth predicate, as in Corollary 3.13.) This would justify us in adding to S as axioms all sentences A for which $\mathbb{A}[\langle A \rangle]$ is derivable in S. Calling the new system S′, we can then repeat this procedure any finite number of times, effectively admitting into S proofs which invoke the release rule a finite number of times. Alternatively, we could avoid the release rule and simply adopt the informal attitude that $\mathbb{A}[\langle A \rangle]$ is "just as good as" A.

The axiom (I) is "internal" in the sense that it does not describe the behavior of $\mathbb{A}$ except by giving a list of sentences which are to be considered valid internally to $\mathbb{A}$. This is a new feature of the predicate interpretation of $\mathbb{A}$ because we have the ability to express in a single axiom the assertibility of a whole family of sentences; if (I) were presented schematically as the family of axioms $\mathbb{A}[\langle A \rangle]$ with A ranging over all axioms of the system besides (I), then every instance would follow from (C).

Our inclusion of the Hilbert axioms in (I) can be used to execute formal derivations internally to $\mathbb{A}$. It is convenient to formulate this idea in terms of Hilbert systems rather than natural deduction because $\mathbb{A}$ is not well suited to handling the temporary assumptions characteristic of the latter.

We use intuitionistic rather than minimal logic because, given that 0 never equals 1, we can vacuously affirm $\ulcorner$whenever $0 = 1$, also $A\urcorner$ for any sentence A, and the same reasoning should work for other interpretations of $\perp$. However, it may be worth noting that the usual intuitionistic justification for the ex falso law is not convincing according to our principles. That justification claims that vacuously every proof of $\perp$ can be converted into a proof of A since there are no proofs of $\perp$, but as we have already discussed, using the claim that there are no proofs of $\perp$ to justify a statement which can then be used in proofs is circular.

The two axioms ($\wedge$) and ($\rightarrow$) enable us to reason under $\mathbb{A}$. This observation allows us to prove the validity of intuitionistic logic in the general setting of any interpreted language. Previously we had shown that classical logic is sound when applied to a modelled language. But since that result involved a truth function for the language, it did not obviously generalize to languages which are not interpreted in a set model. The following lemma shows how we can express the correctness of intuitionistic logic more generally by using assertibility.

Lemma 4.1. *(Assertible soundness of intuitionistic logic) Any sentence which can be derived using intuitionistic logic is assertible. That is,*

$$\mathrm{Th}[\langle A \rangle] \rightarrow \mathbb{A}[\langle A \rangle],$$

where A ranges over all sentences and the predicate Th *is defined to hold of precisely those sentences which are intuitionistically derivable.*

Proof. Suppose some sentence is derivable from the Hilbert axioms. We prove it is assertible by induction on the length of this derivation.

Each step in the derivation is either an axiom or follows from two previous steps by universal modus ponens. In the former case, the conclusion follows from axiom (I). In the latter case, the sentence being derived has the form $\overline{B}$ and is a consequence of two previous steps of the form $\overline{A}$ and $\overline{A \rightarrow B}$; we then have $\mathbb{A}[\langle \overline{A} \rangle]$ and $\mathbb{A}[\langle \overline{A \rightarrow B} \rangle]$ by the induction hypothesis, and $\mathbb{A}[\langle \overline{B} \rangle]$ follows from axiom ($\rightarrow$). This completes the proof. $\square$

This lemma has little content, or at least it should not be very surprising, given that the assertibility of the Hilbert axioms is directly encoded in our axioms for $\mathbb{A}$. The point is not so much the technical content of Lemma 4.1, but its formulation, which shows us how to express the validity of intuitionistic logic for languages which are not interpreted in a set model.

Theorem 4.2. *Let* $A_1, \ldots, A_n$ *and* B *be sentences and suppose the implication*

$$A_1 \wedge \cdots \wedge A_n \to B$$

is intuitionistically derivable. Then

$$\mathbb{A}[\langle A_1 \rangle] \wedge \cdots \wedge \mathbb{A}[\langle A_n \rangle] \to \mathbb{A}[\langle B \rangle].$$

Proof.   First, repeated application of $(\wedge)$ yields

$$\mathbb{A}[\langle A_1 \rangle] \wedge \cdots \wedge \mathbb{A}[\langle A_n \rangle] \to \mathbb{A}[\langle A_1 \wedge \cdots \wedge A_n \rangle].$$

Next, we have

$$\mathbb{A}[\langle A_1 \wedge \cdots \wedge A_n \to B \rangle]$$

by the lemma, and

$$\mathbb{A}[\langle A_1 \wedge \cdots \wedge A_n \rangle] \to \mathbb{A}[\langle B \rangle]$$

follows from this by axiom $(\to)$. Putting this together with the first displayed implication yields the desired result. $\square$

Now we can determine how the assertibility predicate $\mathbb{A}$ interacts with the logical symbols. The next corollary follows easily from Theorem 4.2.

Corollary 4.3. *Let* A *and* B *be sentences and let* $C[x]$ *be a formula with one free variable* x. *Then*

(a) $\mathbb{A}[\langle A \wedge B \rangle] \leftrightarrow \mathbb{A}[\langle A \rangle] \wedge \mathbb{A}[\langle B \rangle]$
(b) $\mathbb{A}[\langle A \rangle] \vee \mathbb{A}[\langle B \rangle] \to \mathbb{A}[\langle A \vee B \rangle]$
(c) $\mathbb{A}[\langle A \to B \rangle] \to (\mathbb{A}[\langle A \rangle] \to \mathbb{A}[\langle B \rangle])$.
(d) $\mathbb{A}[\langle (\forall x) C[x] \rangle] \to \mathbb{A}[\langle C[t] \rangle]$
(e) $\mathbb{A}[\langle C[t] \rangle] \to \mathbb{A}[\langle (\exists x) C[x] \rangle]$

for any constant term t.

If $\mathbb{A}$ is treated as a logical operator then Theorem 4.2 and Corollary 4.3 can be presented in a schematic form which applies to arbitrary formulas, not just sentences. In this case the appeal to Lemma 4.1 in the proof of Theorem 4.2 could be replaced by the capture scheme.

Except for part (a), the above are all implications in one direction only. Going in the opposite direction is more complicated, but this can be achieved by mimicking the corresponding rules of natural deduction. Thus we have the inferences

$$\mathbb{A}[\langle A \vee B \rangle] \wedge \mathbb{A}[\langle A \to C \rangle] \wedge \mathbb{A}[\langle B \to C \rangle] \to \mathbb{A}[\langle C \rangle],$$

which eliminates $\vee$, and

$$\mathbb{A}[\langle (\exists x) A[x] \rangle] \wedge \mathbb{A}[\langle (\forall x)(A[x] \to B) \rangle] \to \mathbb{A}[\langle B \rangle],$$

which eliminates $\exists$. The reverse directions for $\to$ and $\forall$ are somewhat more complicated to state. We can still follow the natural deduction formulations, but now we need to take account of the possible appearance of free variables and the possible assumption of premises. Thus, for example, the introduction rule for $\to$ states that if the formula B is derivable from the premises A and $C_1, \ldots, C_n$ then the formula $A \to B$ is derivable from the premises $C_1, \ldots, C_n$. This translates to the introduction law

$$\mathbb{A}[\langle \overline{A \wedge C_1 \wedge \cdots \wedge C_n \to B} \rangle] \to \mathbb{A}[\langle \overline{C_1 \wedge \cdots \wedge C_n \to (A \to B)} \rangle].$$

The introduction law for $\forall$ would be

$$\mathbb{A}[\langle \overline{C_1 \wedge \cdots \wedge C_n \to A[x]} \rangle] \to \mathbb{A}[\langle \overline{C_1 \wedge \cdots \wedge C_n \to (\forall x) A[x]} \rangle],$$

under the hypothesis that x does not appear freely in $C_1 \wedge \cdots \wedge C_n$.

Our apparent inability to affirm the obvious reverse implications such as $\mathbb{A}[\langle A \vee B \rangle] \to \mathbb{A}[\langle A \rangle] \vee \mathbb{A}[\langle B \rangle]$ or $\mathbb{A}[\langle (\exists x) A[x] \rangle] \to (\exists x) \mathbb{A}[\langle A[\hat{x}] \rangle]$ is a consequence of our general apparent inability to affirm the soundness of $\mathbb{A}$. In classical logic we would not expect these implications: being able to prove $A \vee B$ does not obviously entail that we can prove either A or B. But if we only accept the law of excluded middle in those instances where it is possible, in principle, to determine which conjunct holds, then it seems like this inference should be valid. This is reflected in the standard intuitionistic idea that a proof of $A \vee B$ should be either a proof of A or a proof of B. If so, then the implication $\mathbb{A}[\langle A \vee B \rangle] \to \mathbb{A}[\langle A \rangle] \vee \mathbb{A}[\langle B \rangle]$ would seem to be valid.

But the intuitionistic idea is problematic. Requiring a proof of $A \vee B$ to actually be either a proof of A or a proof of B is surely too strong: we can prove that $2^{11} - 1$ is either prime or composite without determining which it is, and perhaps even without having any practical means of doing so. Intuitionists are well aware of this objection and sometimes explicitly widen the characterization of a proof of a disjunction to allow arguments which merely show us how to determine which of A and B is true. But this brings us back to the familiar equivocation between $A \vee B$ and $\mathbb{A}(A \vee B)$: to know that one can prove $A \vee B$ is to know $\mathbb{A}(A \vee B)$, not to know $A \vee B$.[30]

Essentially the same problem appears, perhaps more clearly, in proofs of $\forall\exists$ sentences. According to some intuitionistic accounts, a proof of $(\exists x)A[x]$ consists of a constant term t together with a proof of $A[t]$. If so, then from the hypothesis $\mathbb{A}[\langle(\exists x)A[x]\rangle]$ we could infer that there is a constant term t such that $\mathbb{A}[\langle A[t]\rangle]$, and this would seem to show that the implication $\mathbb{A}[\langle(\exists x)A[x]\rangle] \rightarrow (\exists x)\mathbb{A}[\langle A[\hat{x}]\rangle]$ is valid. But recall that in order to prove, say, that for every number x there is a largest prime p less than x, it is not enough to show how to generate, for each x, a proof of that instance of the claim. As we already discussed at length, this would only show that for every x it is assertible that such a prime exists. A proof of the stated claim would have to consist of a single argument which, for every x, proves that there is a largest prime p less than $\hat{x}$. But this single argument obviously cannot explicitly present a term which in every case evaluates to the desired prime p, as there is no single term which can do this. The best we can do in proving the general statement is to say how to find p. Thus, we cannot demand that a proof of $(\exists x)A[x]$ explicitly include a constant term which names a suitable value of x, as intuitionistic accounts often require. We have to allow proofs which tell us how to find the desired value, and this raises a soundness issue: we can only infer $(\exists x)\mathbb{A}[\langle A[\hat{x}]\rangle]$ from $\mathbb{A}[\langle(\exists x)A[x]\rangle]$ if the proof of $(\exists x)A[x]$ is sound. We need to know that any proof which purports to direct us to a suitable value of x actually does so. This shows why the implication $\mathbb{A}[\langle(\exists x)A[x]\rangle] \rightarrow (\exists x)\mathbb{A}[\langle A[\hat{x}]\rangle]$ is problematic: it requires a global affirmation of the soundness of existence proofs.

In contrast, the implication $(\forall x)\mathbb{A}[\langle A[\hat{x}]\rangle] \rightarrow \mathbb{A}[\langle(\forall x)A[x]\rangle]$ is initially implausible. Merely assuming that each $A[\hat{x}]$ can be proven individually does not give us any obvious reason to suppose that the single assertion $(\forall x)A[x]$ can be proven. However, if x ranges over a sufficiently small domain then the implication is clearly valid. For instance, the $(\wedge)$ law effectively affirms this implication in the case that x ranges over a set with two elements. In that case, it is clear that we can simply put a proof of A and a proof of B together to get a proof of $A \wedge B$. Certainly the same can be said for any finite family of sentences, and depending on one's views about infinite constructions, it may be possible to go even further in collecting proofs of all instances of $A[\hat{x}]$ to get a single proof of $(\forall x)A[x]$. Thus we can include as an optional additional axiom the following analog of $(\wedge)$:

$$(\forall) \qquad (\forall x)\mathbb{A}[\langle \overline{A[\hat{x}]} \rangle] \rightarrow \mathbb{A}[\langle \overline{(\forall x)A[x]} \rangle]$$

It will be valid in settings where the variable x has sufficiently limited range that proofs of all instances of $A[\hat{x}]$ could be collected together to get a single proof of $(\forall x)A[x]$. I will return to this point in Section 6.3.

The logical operator version of $(\forall)$ would be the scheme

$$(\forall_{lo}) \qquad (\forall x)\mathbb{A}(A[x]) \rightarrow \mathbb{A}((\forall x)A[x])$$

with A ranging over all formulas.

4.6 Applications

Summary of problems with local truth. Definition of a truth predicate for an interpreted language. Propositions, concepts, and relations. Second-order logic.

In Section 2.4 I summarized the array of difficulties we had encountered involving propositions, concepts, relations, and truth predicates for interpreted languages. These all seem like meaningful ideas, but we apparently cannot define them rigorously without invoking a global notion of classical truth. And as the liar paradox shows, that is something we do not have.

We do have local notions of classical truth; this was shown by example in arithmetical settings in Sections 1.4 — 1.6, and more generally for languages with set models in Section 2.3. This partially solves the problem, because once we have a truth predicate for arithmetic, say, we can specify the identity criteria for things like propositions and concepts in the context of the language of arithmetic with no further difficulty. For many purposes this partial solution may be perfectly adequate. However, it is fundamentally unsatisfactory for several reasons.

First, we sometimes need to use interpreted languages (in the general sense discussed in Section 2.3) whose interpretation does not come from a set model. The most obvious example of this is when we want to make assertions about all sets. For instance, every set is a subset of itself: in order to say this we need a variable that ranges over all sets, so we must not be using a language that is interpreted in a set model. It needs to have a class model so that variables can range over the class of all sets. But once classes are part of the picture we have to be able to talk about them too, and this cannot be done even with a class model. At this point it should be clear that the local approach afforded by the truth functions we can define for languages interpreted in set models is inadequate in general.

A second problem for the local truth approach is our apparent inability to say what constitutes a truth predicate for an interpreted language. We can give specific constructions, but spelling out the condition these constructions satisfy that makes them truth predicates seems to require the use of a broader truth predicate than the ones under discussion. So without appealing to a fictional global truth predicate we can never get beyond the limited settings in which we can already express schematic assertions without using truth, and in which truth predicates are therefore superfluous. We analyzed this issue at some length in Chapter 2 but found no good solution there.

The third problem is the one discussed in Section 2.7: local truth predicates are unsuited to the task of quantifying over second-order variables. We can make global statements about all sets by passing to a language with a class model, and we can make global statements about all classes by passing to an interpreted language in the general sense of Section 2.3, but there does not seem to be any way at all, in any language, to say that for an arbitrary concept C, if $\emptyset$ falls under C then $\emptyset$ falls under C. Since we need a notion of truth to have a notion of falling under, the best we can do at this point is to make this assertion of all concepts in some target language. This assertion would therefore have to be expressed in a language which could also express concepts to which it did not apply. Thus, the general fact about all concepts appears to be another example of something that cannot be said in any language, which I have insisted is not an acceptable state of affairs.

But now these difficulties can all be resolved. We can say that T is a classical truth predicate for an interpreted language $\mathbb{L}$ if for every sentence A of $\mathbb{L}$ the equivalence $\ulcorner T(A) \leftrightarrow \hat{A} \urcorner$ is assertible, where $\hat{A}$ is the metalanguage translation of A determined by the interpretation. This is how we resolve the Tarskian catastrophe: instead of requiring each instance of the T-scheme to be true, which assumes a global notion of truth that does not exist, or to be derivable in some formal metasystem, which fails to capture basic properties of truth, we require each instance to be assertible —

provable in a general semantic sense.[31]

Finally we have the means to express the fact that "$\tau(A) = 1$", with free variable A, is a truth predicate in the constructions of Sections 1.4 — 1.6, or more generally in the construction of Section 2.3. In each case, for every sentence A in the language under consideration, the equivalence $\ulcorner \tau(\text{"}A\text{"}) = 1 \leftrightarrow \hat{A} \urcorner$ is assertible. According to the definition we just gave, this is exactly what it means for "$\tau(A) = 1$" to be a truth predicate. And it is legitimate because we showed, in those earlier sections, that we could prove this equivalence for every sentence A; so each of the desired equivalences is indeed assertible.

This conclusion is especially satisfying because one of the frustrating aspects of that earlier discussion was our inability to affirm that all of these equivalences were true when we knew that we could prove them. It turns out that knowing we can prove each sentence in some family simply licenses us to say that every sentence in the family can be proven, i.e., is assertible. There is a possible explanation here of why we have such a strong native intuition for the idea that there must be a global notion of truth. Perhaps what we really have an intuition for is not the existence of a concept that performs a global disquotational function, but the existence of a global concept of rational justifiability.

This may be a good place to recall that in constructive mathematics "truth" is understood as assertibility. Of course in constructive mathematics it is generally assumed that the assertibility predicate satisfies the T-scheme, which would actually make it a classical truth predicate. But there could also be a sense among intuitionists that classical truth has some kind of dubious metaphysical substance which ought to be replaced by the more concretely accessible notion of assertibility. The first part of that statement is wrong — classical truth has no deep metaphysical substance; classical truth predicates are nothing more than ordinary predicates which perform a disquotational function, and as such they should cause no offense to constructivist scruples — but the idea that assertibility is a form of truth might very well capture an important source of confusion about truth among nonconstructivists.

It might also capture an aspect of natural language. Kripke points out that in ordinary speech truth can be used in a circular way that has the potential to be either meaningful or paradoxical.[32] (Jones: "Most of Nixon's assertions about Watergate are false". Nixon: "Everything Jones says about Watergate is true".) In such cases Tarski's crude conclusion that natural language is inconsistent does not seem very helpful, but on the other hand, Kripke's theory of a partial truth predicate defined as a fixed point of a transfinite recursive construction seems rather far from ordinary usage. We do not even have a term for "grounded truth" in nonphilosophical language. The basic intuition for grounded truth, that a sentence is groundedly true if one is eventually able to chase down a positive truth value, might work better as an intuition for assertibility in the broad sense of being knowable in principle.

Our definition of truth predicates is not so far from Tarski's: we simply replace derivability in some metasystem with assertibility. Does it, then, suffer from the same defects we identified in Section 3.7? No. First, we have a single notion of what counts as a truth predicate, not one which varies from metasystem to metasystem. Second, on our definition truth predicates have additional structure. For instance, according to Lemma 4.1, for every sentence A the implication $\ulcorner A \to A \urcorner$ is assertible. It easily follows that if T is a truth predicate in our sense then for every sentence A the sentence $T(\ulcorner A \to A \urcorner)$ is assertible, and if one has sufficient infinitary resources, one can go on to infer that $(\forall A)T(\ulcorner A \to A \urcorner)$ is assertible using the $(\forall)$ law. (See Section 6.3.) Thus, under our characterization of classical truth predicates the law $\ulcorner A \to A \urcorner$ is a general feature of truth.

Now that we know what constitutes a truth predicate for an interpreted language, we can also give a general account of what it means for two sentences to express the same proposition. If T is a truth predicate for a language then we can take $T(\ulcorner A \leftrightarrow B \urcorner)$ as the condition under which two sentences A and B in the language express the same proposition. This definition seems to depend on the choice of truth predicate, but we can easily prove that if T' is any other truth predicate for the language then for all A and B the equivalence

$$T(\ulcorner A \leftrightarrow B \urcorner) \qquad \leftrightarrow \qquad T'(\ulcorner A \leftrightarrow B \urcorner)$$

is assertible. However, it might be simpler to avoid truth predicates altogether, and just say that A and B express the same proposition if $\ulcorner A \leftrightarrow B \urcorner$ is assertible. Similarly, we can declare that C and D express the same concept if $\ulcorner (\forall x)(C[[x]] \leftrightarrow D[[x]]) \urcorner$ is assertible, and so on.

Thus, when we specify the identity criteria for propositions, concepts, and relations, we have a choice between either giving a local condition that uses a classical truth predicate for some target language or giving a global condition that uses assertibility. The definition can be a local classical one or a global constructive one. The global characterization of the local classical definitions is constructive — we need to use assertibility to say what constitutes a classical truth predicate in a general setting — but in any particular case, once we have settled on a local classical truth predicate the constructive aspect is invisible. The point is that local classical truth predicates apply to target languages from some external perspective, so that there is no self-referential aspect. In those cases where we need to talk within a language $\mathbb{L}$ about, say, all concepts expressible in $\mathbb{L}$, the classical option is not viable.

For example, as we discussed in Section 2.7, classical truth predicates cannot help us express global second-order facts, because classical truth predicates are by definition local. In this case the direct use of assertibility is our only option. Thus, the way to globally express that every sentence implies itself is by saying that for every sentence A, the implication $\ulcorner A \rightarrow A \urcorner$ is assertible. Indeed, we are entitled to say this by Lemma 4.1. Similarly, we can define a global falling under relation by saying that x falls under C if $C[[x]]$ is assertible. And we can then legitimately say that for any concept C, if $\emptyset$ falls under C then $\emptyset$ falls under C. That is equivalent to the trivial claim $(\forall C)(\mathbb{A}[C[[\emptyset]]] \rightarrow \mathbb{A}[C[[\emptyset]]])$. What was not trivial was establishing a legitimate way to say this. But we can also make more substantive statements, such as: for any concept C, if $C[[\hat{0}]]$ is assertible and for every n the implication $C[[\hat{n}]] \rightarrow C[[\hat{n}']]$ is assertible, then for every n the sentence $C[[\hat{n}]]$ is assertible.

The preceding discussion will be made more rigorous in Sections 5.5 and 5.6, when we discuss formal systems for reasoning about concepts using the global falling under relation defined in terms of assertibility.

Chapter 5

Systems

5.1 Pure assertibility

The system PAR. *It is completely consistent. Is assertibility reasoning consistent in general?*

We now have some idea of how to reason about the concept of assertibility. The most important question to answer is whether the axioms for $\mathbb{A}$ presented in Chapter 4 are consistent. A secondary question is, on the positive side, what kinds of things can we legitimately prove about seemingly paradoxical sentences such as the assertible liar sentence?

In this chapter I will present a small variety of formal systems for reasoning about assertibility in various settings and show that they are consistent in a strong sense.[1] Let us start with a very minimal setting in which there are no variables and the only subject of discussion is the assertibility of the sentences of the language itself. Such a setting could include a liar sentence which states of itself that it is not assertible, or a variant liar sentence which states that its negation is assertible. It could include pairs of sentences which comment on each other's assertibility, and so on.

We can formalize this setting using the following language. First, we have a finite list of constant symbols $\dot{\perp}$, L_1, ..., L_n. Each constant symbol is a constant term, and whenever t_1 and t_2 are constant terms so are $(t_1 \dot{\wedge} t_2)$, $(t_1 \dot{\vee} t_2)$, $(t_1 \dot{\rightarrow} t_2)$, and $\dot{\mathbb{A}}[t_1]$. That completely describes the constant terms. The atomic sentences of the language are $\perp$ and $\mathbb{A}[t]$ for any constant term t, and whenever A and B are sentences so are $(A \wedge B)$, $(A \vee B)$, and $(A \rightarrow B)$. That completely describes the sentences.

We will regard $\dot{\perp}$ as a *name* for $\perp$ and the L_i as names for specific sentences of the language. For concreteness, we can take $n = 4$ and let L_1, L_2, L_3, and L_4 respectively name the sentences $\neg\mathbb{A}[L_1]$, $\mathbb{A}[\dot{\neg}L_2]$, $\neg\mathbb{A}[L_4]$,

and $\mathbb{A}[L_3]$. (As usual, $\neg A$ abbreviates $(A \to \bot)$ and $\dot\neg t$ abbreviates $(t \dot\to \dot\bot)$.)
In ordinary language we could express these sentences as follows:

L_1: "The sentence L_1 is not assertible."
L_2: "The negation of L_2 is assertible."
L_3: "The sentence L_4 is not assertible."
L_4: "The sentence L_3 is assertible."

Thus L_1 is an assertible liar sentence, L_2 is a variant assertible liar sentence, and L_3 and L_4 constitute an assertible liar pair.

Recursively, if the terms t_1 and t_2 respectively name the sentences A and B, then $(t_1 \dot\wedge t_2)$ names the sentence $(A \wedge B)$, $(t_1 \dot\vee t_2)$ names the sentence $(A \vee B)$, $(t_1 \dot\to t_2)$ names the sentence $(A \to B)$, and $\dot{\mathbb{A}}[t_1]$ names the sentence $\mathbb{A}[t_1]$. Thus, every term names a unique sentence.

We may adopt the following schematic versions of the axioms (C), ($\wedge$), and ($\to$) from Section 4.5:

(C) $A \to \mathbb{A}[t_1]$
($\wedge$) $\mathbb{A}[t_1] \wedge \mathbb{A}[t_2] \to \mathbb{A}[(t_1 \dot\wedge t_2)]$
($\to$) $\mathbb{A}[t_1] \wedge \mathbb{A}[(t_1 \dot\to t_2)] \to \mathbb{A}[t_2]$

for all terms t_1 and t_2, where A is the sentence named by t_1. I call the system with the above axioms and intuitionistic logic *Pure Assertibility Reasoning (PAR)* for the sentences L_i. For the sake of brevity we can simply call the system PAR, suppressing its dependence on n and the choice of the sentences named by the L_i.

Since there are no variables in the language, we have to express the axioms ($\wedge$) and ($\to$) as schemes. There is no need for a schematic version of (I) because these individual instances are consequences of the capture scheme. Obviously PAR describes a very limited type of reasoning. However, interesting results can still be obtained. For example, with respect to the L_i specified above, since L_1 is a name for $\neg\mathbb{A}[L_1]$, capture yields $\neg\mathbb{A}[L_1] \to \mathbb{A}[L_1]$, from which we can infer $\neg\neg\mathbb{A}[L_1]$. Thus, we have derived the sentence

$$\dot\neg L_1,$$

the negation of the assertible liar sentence. We have to be a little careful here; "$\neg L_1$" is not a sentence of the language, but "$\dot\neg L_1$" is a term which names the sentence $\neg\neg\mathbb{A}[L_1]$. The version

$$\neg\neg\mathbb{A}[L_1]$$

can be read as stating that the assertible liar sentence is not not assertible. (But in the absence of double negation elimination, this does not entail that the assertible liar sentence is assertible.) Next, applying capture to $\dot{\neg}L_1$ yields $\mathbb{A}[\dot{\neg}L_1]$. So if we assumed that L_1 is assertible we would have both $\mathbb{A}[L_1]$ and $\mathbb{A}[\dot{\neg}L_1]$, and Theorem 4.2 would then yield $\mathbb{A}[\dot{\bot}]$. This shows that

$$\mathbb{A}[L_1] \to \mathbb{A}[\dot{\bot}].$$

Entailing that a contradiction is provable is not quite as bad as entailing a contradiction outright, but it still intuitively expresses that the premise is wrong. We may say that $\dot{\mathbb{A}}[L_1]$ is *weakly false*.

Assume the law of assertible noncontradiction, $\mathbb{A}[\dot{\bot}] \to \bot$. Since we just showed $\mathbb{A}[L_1] \to \mathbb{A}[\dot{\bot}]$, under this new hypothesis we can deduce $\mathbb{A}[L_1] \to \bot$, i.e., $\neg\mathbb{A}[L_1]$. But earlier we derived $\neg\neg\mathbb{A}[L_1]$, so this yields a contradiction. We conclude that $(\mathbb{A}[\dot{\bot}] \to \bot) \to \bot$, i.e., we have derived that

$$\neg\neg\mathbb{A}[\dot{\bot}].$$

The law of assertible noncontradiction is false. That is not to say that $\bot$ is assertible, only that it is not not assertible. Indeed, since $\bot$ implies any sentence, we can now use Theorem 4.2 to infer $\neg\neg\mathbb{A}[t]$ for any term t in the language. This reflects our inability to affirm that any sentence cannot be proven. Finally, $\neg\neg\mathbb{A}[L_1]$ — equivalently, $\neg\mathbb{A}[L_1] \to \bot$ — together with the capture law $\bot \to \mathbb{A}[\dot{\bot}]$ yields $\neg\mathbb{A}[L_1] \to \mathbb{A}[\dot{\bot}]$. Since we already showed $\mathbb{A}[L_1] \to \mathbb{A}[\dot{\bot}]$, we may conclude that

$$(\mathbb{A}[L_1] \vee \neg\mathbb{A}[L_1]) \to \mathbb{A}[\dot{\bot}]$$

as well. This instance of the law of excluded middle is weakly false.

So it is not as if we cannot say anything about paradoxical sentences. To the contrary, we can derive a range of interesting and substantive conclusions about them. The assertible liar sentence is false and its assertibility is weakly false, for example. The law of assertible noncontradiction is false and an instance of the law of excluded middle is weakly false.

The variant assertible liar sentence L_2 has slightly different, and slightly nicer, properties than the standard assertible liar sentence. Using reasoning similar to that used on L_1, we can show that L_2 is weakly false, its negation is false, and excluded middle weakly fails for $L_2 \dot{\vee} \dot{\neg}L_2$. That is, we can derive the sentences $L_2 \dot{\to} \dot{\mathbb{A}}[\dot{\bot}]$, $\dot{\neg}\dot{\neg}L_2$, and $(L_2 \dot{\vee} \dot{\neg}L_2) \dot{\to} \dot{\mathbb{A}}[\dot{\bot}]$. As for the assertible liar pair, L_3 is false and L_4 is weakly false.

But is this system consistent? Yes, it is easy to see that $\perp$ is not derivable in PAR. Assign classical truth values to the sentences of the language by setting $\tau(\perp) = 0$ and $\tau(\mathbb{A}[t]) = 1$ for all t, and evaluating τ on complex sentences in the usual recursive way. Then it is easy to see that τ takes the value 1 on all the axioms and that formal derivations preserve this property. (Most straightforwardly, we could do this by reasoning using Hilbert-style derivations. One checks that τ takes the value 1 on the Hilbert axioms (1) — (9), and it is immediate from the definition of $\tau(A \to B)$ that if $\tau(A) = \tau(A \to B) = 1$ then $\tau(B) = 1$.)

A more interesting question is whether $\mathbb{A}[\dot{\perp}]$ can be derived. If none of $\perp$, $\mathbb{A}[\dot{\perp}]$, $\mathbb{A}^2[\dot{\perp}] = \mathbb{A}[\dot{\mathbb{A}}[\dot{\perp}]]$, ... are derivable in some formal system S then I will say that S is *completely consistent*. This is equivalent to the consistency of the system obtained by augmenting S with the release rule discussed in Section 4.5.

Showing the complete consistency of PAR requires something more than a classical truth function. We know this because we saw above that $\neg\neg\mathbb{A}[\dot{\perp}]$ may be derivable in the system, and any classical truth function that makes $\neg\neg\mathbb{A}[\dot{\perp}]$ true would also make $\mathbb{A}[\dot{\perp}]$ true.

Theorem 5.1. PAR *is completely consistent.*

Proof.   Recursively define a sequence $\mathcal{F}_0$, $\mathcal{F}_1$, $\mathcal{F}_2$, ... of sets of sentences according to the following prescription. $\mathcal{F}_0$ contains $\perp$ but no other atomic sentence; for $n \geq 1$ we let $\mathcal{F}_n$ contain both $\perp$ and every $\mathbb{A}[t]$ such that the sentence named by t belongs to $\mathcal{F}_{n-1}$. For any $n \geq 0$ complex sentences are included in $\mathcal{F}_n$ according to the rules

(1) $A \wedge B$ belongs to $\mathcal{F}_n$ if either A belongs to $\mathcal{F}_n$ or B belongs to $\mathcal{F}_n$ (or both)
(2) $A \vee B$ belongs to $\mathcal{F}_n$ if both A and B belong to $\mathcal{F}_n$
(3) $A \to B$ belongs to $\mathcal{F}_n$ if there exists $k \leq n$ such that A does not belong to $\mathcal{F}_k$ but B does belong to $\mathcal{F}_k$.

We show by induction on n that $\mathcal{F}_n \subseteq \mathcal{F}_{n+1}$ for all n. It will suffice to show that every atomic sentence in $\mathcal{F}_n$ also belongs to $\mathcal{F}_{n+1}$, as $\mathcal{F}_n \subseteq \mathcal{F}_{n+1}$ then follows by induction on complexity. But in the base case $n = 0$ this is trivial because the only atomic sentence in $\mathcal{F}_0$ is $\perp$, and at higher levels, once we have that $\mathcal{F}_n \subseteq \mathcal{F}_{n+1}$, it follows immediately from the rule for inclusion of atomic sentences that every atomic sentence in $\mathcal{F}_{n+1}$ also belongs to $\mathcal{F}_{n+2}$. So the induction can proceed.

Let $\mathcal{F} = \bigcup \mathcal{F}_n$. Then it is straightforward to verify that the complementary set $\mathcal{F}^c$ of sentences that do not belong to any $\mathcal{F}_n$ contains all the axioms of PAR and all the propositional Hilbert axioms (i.e., the axiom schemes (1) — (9) from Section 3.2), and is stable under modus ponens. For example, if there exists n such that A does not belong to $\mathcal{F}_n$ but B does belong to $\mathcal{F}_n$, then $A \to B$ must belong to $\mathcal{F}_n$; it follows that if neither A nor $A \to B$ belong to $\mathcal{F}$ then neither does B. That is, $\mathcal{F}^c$ is stable under modus ponens. All the other claims are at a similar level of difficulty. However, $\bot \in \mathcal{F}_0$, and inductively $\mathbb{A}^n[\dot{\bot}] \in \mathcal{F}_n$ for $n \geq 1$. So we conclude that none of these sentences is derivable in PAR. $\qquad\square$

Can we conclude that our conception of assertibility is consistent? This question is a bit subtle because, as we discussed in detail in Section 4.4, we have not accepted that $\bot$ is not assertible. Symbolically, we are not assuming $\neg\mathbb{A}[\dot{\bot}]$, the law of assertible noncontradiction. We do not rule out the possibility that there could be a valid argument that reaches a false conclusion. As unlikely as that sounds, the assertible liar paradox provides us with a plausible candidate for such a thing.

Accepting paradoxical reasoning as valid would be disastrous, as the ex falso law would then allow us to infer any conclusion at all. But the good news is that we have not reached that point yet. As long as we have doubts about the release law, paradoxical reasoning is blocked. So we may provisionally trust that logical reasoning is sound, with the caveat that the moment we affirm this as an established fact, we will have enabled the assertible liar paradox and will be able to prove a falsehood.

Thus, I do not aim and would not want to show that $\bot$ cannot be proven, in the general semantic sense of "proven". My concern here is merely to show that it cannot be derived from the axioms for $\mathbb{A}$ given in Chapter 4. But I explicitly leave open the possibility that new axioms could be accepted as correct which would make assertibility reasoning inconsistent — in particular, $\neg\mathbb{A}[\dot{\bot}]$ could be accepted as correct.

5.2   Arithmetical assertibility

The system PA$^{\mathbb{A}}$. *It verifies its own soundness and consistency. It is completely consistent.*

The system PAR presented in the last section allows for basic reasoning about assertibility, but as a propositional system it lacks the ability to express general laws. In order to accomplish this we need to be able to quantify over sentences, and one way to do that is to work in an arithmetical setting in which a system of Gödel numbering is available.

When Peano Arithmetic is augmented with an assertibility predicate, our expressive power increases dramatically. This setting exhibits the same self-referential features as PAR: the obvious analog of the Gödel sentence used in Theorem 3.7 (or its Tarskian version used in Theorem 3.8) effectively states of itself that it is not assertible, and with a little more work one can create assertible liar pairs which comment on each other's assertibility, or indeed any finite network of assertibility claims that could be expressed in PAR. This shows that the complete consistency result for PAR, Theorem 5.1, is subsumed by Theorem 5.3 below.

We define the augmented arithmetical system PA^{A} as follows. Its language is the language of arithmetic with one additional concept symbol A. The formula $A[x]$ is interpreted as expressing that x is the Gödel number of an assertible sentence, where we have fixed a Gödel numbering of the language of PA^{A}. Thus the assertibility predicate is self-applicative. Since we have to reason intuitionistically in the presence of a self-applicative assertibility predicate, PA^{A} will use intuitionistic logic. We can accomodate the classical nature of the natural numbers by including as nonlogical axioms the universal closures of all instances of the law of excluded middle $A \vee \neg A$ where A is a formula of pure arithmetic (i.e., it does not contain the symbol "A").

The justification for this inclusion depends on an acceptance of countably long constructions. Recall that assertibility reasoning demands that we reject the existence of unknowable truth values. Thus we cannot justify the law of excluded middle for arithmetical sentences by simply postulating that all such sentences have well-defined truth values; we have to commit ourselves to the idea that these truth values could, in principle, be mechanically evaluated by countably long computations. If countable constructions are available then we can mechanically evaluate the truth function τ on any sentence in the language of arithmetic and can directly verify $A \vee \neg A$ for any formula A, for any assignment of values to its free variables. In the present section we therefore adopt the position that countable constructions are legitimate.

A separate consequence of this position is the validity of the law $(\forall)$ from Section 4.5 for variables which range over the natural numbers. If $A[\hat{x}]$ is eventually accepted as proven, separately for each numerical value of x, then — granting that countably long processes can be run to completion

— there will eventually come a point at which all instances of $A[\hat{x}]$ have been proven. Thus $(\forall x)\mathbb{A}[\langle A[\hat{x}]\rangle] \to \mathbb{A}[\langle(\forall x)A[x]\rangle]$. (See Section 6.3 for more on this point.)

The appropriate axioms for $\text{PA}^\mathbb{A}$ are therefore the universal closures of

- the Hilbert axioms for intuitionistic logic in the language of $\text{PA}^\mathbb{A}$, plus the law of excluded middle for all arithmetical formulas
- the Peano axioms (p. 70), including the induction scheme (10) for all formulas in the language of $\text{PA}^\mathbb{A}$
- the axiom scheme (C) and the axioms $(\wedge)$, $(\to)$, (I), and $(\forall)$ from Section 4.5.

The Hilbert axioms were listed in Section 3.2. Here axiom (I) takes the form $Ax[x] \to \mathbb{A}[x]$ where $Ax[x]$ is an arithmetical formula which expresses that x is the Gödel number of one of the other axioms of $\text{PA}^\mathbb{A}$ (i.e., besides (I) itself). We can assume that for each value of x for which $Ax[\hat{x}]$ holds, this fact can be derived in PA.

Recall the notation $\mathbb{D}_S[x]$ from Section 3.4, which represents an arithmetical expression of the statement that x is the Gödel number of a sentence derivable in S, and the notation $\text{Con}(S)$ from Section 3.5, which represents an arithmetical expression of the statement that S is consistent, i.e., it does not derive $\bot$. We will now show that $\text{PA}^\mathbb{A}$ verifies its own soundness in the sense that it derives a single statement which affirms that every sentence derivable in $\text{PA}^\mathbb{A}$ is assertible. It also derives the assertibility of its own consistency.

Theorem 5.2. *The sentences*

$$(\forall x)(\mathbb{D}_{\text{PA}^\mathbb{A}}[x] \to \mathbb{A}[x])$$

(assertible soundness) and

$$\mathbb{A}[\langle\text{Con}(\text{PA}^\mathbb{A})\rangle]$$

(assertible consistency) are derivable in $\text{PA}^\mathbb{A}$.

Proof. Working in $\text{PA}^\mathbb{A}$, we know from the internal axiom that we have $\mathbb{A}[x]$ whenever x is the Gödel number of any of the other axioms. Also, if I is the internal axiom itself then we can use capture to infer $\mathbb{A}[\langle I\rangle]$. So we can derive, in $\text{PA}^\mathbb{A}$, a single sentence which affirms $\mathbb{A}[x]$ for any x which is the Gödel number of any of the axioms of $\text{PA}^\mathbb{A}$.

Still working in $\text{PA}^\mathbb{A}$, we proceed to verify $\mathbb{A}[x]$ whenever x is the Gödel number of any sentence that is derivable in $\text{PA}^\mathbb{A}$. The argument goes by induction on the length of the derivation. Given the result of the preceding paragraph, and since we are reasoning using Hilbert-style deduction with universal modus ponens as the only rule of inference, the desired conclusion immediately follows from the ($\rightarrow$) axiom. (Essentially the same argument was made in the proof of Lemma 4.1.) This completes the proof of assertible soundness.

For assertible consistency, I first claim that we can derive, in $\text{PA}^{\mathbb{A}}$, the sentence

$$(\forall x)\mathbb{D}_{\text{PA}^{\mathbb{A}}}[\langle\neg\text{Der}_{\text{PA}^{\mathbb{A}}}[\langle\bot\rangle, \hat{x}]\rangle],$$

which affirms that for every number x the system $\text{PA}^{\mathbb{A}}$ derives that x is not the Gödel number of a dervation in $\text{PA}^{\mathbb{A}}$ of $\bot$. In light of the second incompleteness theorem this may seem surprising. The proof goes as follows. Reasoning in $\text{PA}^{\mathbb{A}}$, we argue that for each x, either x is not the Gödel number of a derivation in $\text{PA}^{\mathbb{A}}$ of $\bot$, in which case this fact is mechanically verifiable and hence derivable in $\text{PA}^{\mathbb{A}}$ (indeed, even in PA), or else x is the Gödel number of a derivation in $\text{PA}^{\mathbb{A}}$ of $\bot$, in which case $\text{PA}^{\mathbb{A}}$ is inconsistent and therefore derives anything. This establishes the claim.[2] By assertible soundness we can infer

$$(\forall x)\mathbb{A}[\langle\neg\text{Der}_{\text{PA}^{\mathbb{A}}}[\langle\bot\rangle, \hat{x}]\rangle],$$

and then using axiom ($\forall$) we can infer $\mathbb{A}[\langle\text{Con}(\text{PA}^{\mathbb{A}})\rangle]$.   □

Theorem 5.2 may be contrasted with Corollary 3.13, a crucial difference being that $\text{PA}^{\mathbb{T}}$ establishes the soundness of PA while $\text{PA}^{\mathbb{A}}$ establishes its own (assertible) soundness.

We will now prove that $\text{PA}^{\mathbb{A}}$ is completely consistent. The proof is a strengthened version of the proof of Theorem 5.1.

Theorem 5.3. *Let A_0 be a false sentence in the language of arithmetic. Then none of A_0, $\mathbb{A}[\langle A_0\rangle]$, $\mathbb{A}^2[\langle A_0\rangle]$, ..., is derivable in $\text{PA}^{\mathbb{A}}$.*

Proof.   We define a sequence $(\mathcal{F}_n)$ such that each $\mathcal{F}_n$ is a set of sentences in the language of $\text{PA}^{\mathbb{A}}$. The definition proceeds by recursion on n, and for a given value of n by recursion on the complexity of a sentence. Fixing n, we define $\mathcal{F}_n$ as follows. The atomic sentences have the form $t_1 = t_2$ and $\mathbb{A}[t_1]$ where t_1 and t_2 are numerical terms (i.e., they contain no variables). (Recall that in arithmetical settings we take "$\bot$" to stand for "$\hat{0} = \hat{1}$", so $\bot$ is not a separate atomic sentence.) Put $t_1 = t_2$ in $\mathcal{F}_n$ if t_1 and t_2 numerically evaluate to different numbers, and for $n \geq 1$ put $\mathbb{A}[t_1]$ in $\mathcal{F}_n$ if

t_1 evaluates to the Gödel number of a sentence that belongs to $\mathcal{F}_{n-1}$. We do not place any sentence of the form $\mathbb{A}[t_1]$ in $\mathcal{F}_0$.

Place $A \wedge B$ in $\mathcal{F}_n$ if either A or B belongs to $\mathcal{F}_n$; place $A \vee B$ in $\mathcal{F}_n$ if both A and B belong to $\mathcal{F}_n$; place $(\forall x)A[x]$ in $\mathcal{F}_n$ if $A[\hat{x}]$ belongs to $\mathcal{F}_n$ for some x; and place $(\exists x)A[x]$ in $\mathcal{F}_n$ if $A[\hat{x}]$ belongs to $\mathcal{F}_n$ for all x. Place $A \to B$ in $\mathcal{F}_n$ if for some $k \le n$ we have $A \notin \mathcal{F}_k$ and $B \in \mathcal{F}_k$.

The proof that $\mathcal{F}_n \subseteq \mathcal{F}_{n+1}$ for all n is basically the same as in the proof of Theorem 5.1. Let $\mathcal{F} = \bigcup \mathcal{F}_n$. It is tedious but straightforward to verify both that no axiom of $\mathrm{PA}^{\mathbb{A}}$ belongs to $\mathcal{F}$ and that the complement of $\mathcal{F}$ is stable under universal modus ponens. Thus no theorem of $\mathrm{PA}^{\mathbb{A}}$ lies in $\mathcal{F}$. But every false arithmetical sentence belongs to $\mathcal{F}_0$ — this can be seen by induction on the complexity of the sentence — so $\mathbb{A}^n[\langle A_0 \rangle]$ belongs to $\mathcal{F}_n$ for every $n \ge 1$. Therefore none of these statements can be derivable in $\mathrm{PA}^{\mathbb{A}}$. $\qquad\square$

Even in the case where A_0 is $\bot$, the soundness of PA, not its mere consistency, is essential to this result.[3] For example, $\mathrm{PA} + \neg\mathrm{Con}(\mathrm{PA})$ is consistent but not sound, and if we augment it with an assertibility predicate the resulting system derives $\mathbb{A}[\langle \bot \rangle]$. This is because it derives $\mathbb{A}[\langle \neg\mathrm{Con}(\mathrm{PA}) \rangle]$ by capture, but it also derives $\mathbb{A}[\langle \mathrm{Con}(\mathrm{PA}) \rangle]$ by Theorem 5.2 (since it is stronger than $\mathrm{PA}^{\mathbb{A}}$ and $\mathrm{Con}(\mathrm{PA}^{\mathbb{A}})$ implies $\mathrm{Con}(\mathrm{PA})$). So $\mathrm{PA} + \neg\mathrm{Con}(\mathrm{PA})$ with assertibility is not completely consistent. (It is still consistent, by a simple modification of the straight consistency proof of PAR mentioned in Section 5.1: find a set model for $\mathrm{PA} + \neg\mathrm{Con}(\mathrm{PA})$ and define a truth function for the augmented system which evaluates arithmetical truth in this model and assigns the truth value 1 to $\mathbb{A}[\hat{x}]$ for every x. Then every derivable sentence has the truth value 1, but $\bot$ has the truth value 0, so the system is consistent.)

Replacing $\mathrm{Con}(\mathrm{PA})$ in the preceding parenthetical argument with any sentence that is not derivable in PA shows that an arithmetical sentence is derivable in $\mathrm{PA}^{\mathbb{A}}$ if and only if it is derivable in PA. However, $\mathrm{PA}^{\mathbb{A}}$ becomes stronger at higher levels of assertibility, which is to say that the assertibility version of Löb's theorem is false. For example:

Corollary 5.4. $\mathrm{PA}^{\mathbb{A}}$ *derives* $\mathbb{A}[\langle \mathrm{Con}(\mathrm{PA}^{\mathbb{A}}) \rangle]$ *but not* $\mathrm{Con}(\mathrm{PA}^{\mathbb{A}})$.

The first part was proven in Theorem 5.2, and the second part follows from Theorem 3.9 since $\mathrm{PA}^{\mathbb{A}}$ is consistent and hence cannot derive its own consistency. (We proved Theorem 3.9 for classical systems, but it holds in the intuitionistic setting as well, by identical reasoning. Or we can avoid applying Theorem 3.9 to PA^A by invoking the observation made above that PA and PA^A derive the same arithmetical sentences, and noting that $Con(PA^A) \rightarrow Con(PA)$, derivably in PA.)

Although PA^A is completely consistent, Theorem 5.2 does not contradict Theorem 3.9. We can derive in PA^A that its Gödel sentence is assertible, or equivalently, we can derive the assertibility of the statement that the Gödel sentence cannot be derived. And indeed the Gödel sentence cannot be derived; only its assertibility can be derived. So this subtle distinction allows us to effectively verify the consistency of the system PA^A within PA^A itself.

5.3   Problems of rational agency

Naturalistic trust. Reflective trust. Reflectively coherent trust. Disjunctive trust. The assertibility rule.

The fact that arithmetical systems can be formulated which effectively affirm their own soundness has a surprising application to an interesting problem in theoretical artificial intelligence.

The inability of a rational agent who reasons within a formal system S to affirm the soundness, or even the consistency, of S is an unhappy but familiar phenomenon. However, in a recent paper Yudkowsky and Herreshoff observe that this phenomenon carries a sharper sting in the context of an AI that is licensed to act under precisely defined conditions.[4] Thus, imagine an intelligent machine M that is capable of reasoning within S and is licensed to perform some action α_0 when it has verified the truth of some sentence A_0, i.e., when it has derived A_0 in S. I will now describe a variety of situations in which M is frustratingly unable to justify performing α_0 even though it intuitively ought to be able to do so.[5]

Naturalistic trust. Suppose M decides to improve its performance by constructing an assistant M' whose job is to derive sentences for M. M programs the assistant to reason within the same system S that M reasons in, and to alert M when it has derived the sentence A_0 (or any other sentence which expresses an actionable criterion for M).

The puzzle is that even though M knows that M' reasons within S, it cannot act on the information that M' has derived A_0. Knowing this tells M only that A_0 is derivable in S, not that A_0 is true, and lacking a soundness scheme it cannot infer A_0 from $\mathbb{D}_S[\langle A_0 \rangle]$. In fact nothing M' can tell M, short of a line by line account of the actual derivation, could convince M that A_0 is true. So apparently the best M can do in this situation is to retrieve the formal derivation of A_0 from M' and check it line by line, thereby establishing A_0 to M's own satisfaction and licensing the action α_0.

This seems wrong because M knows perfectly well what the outcome of this verification is going to be, but still has no way to get around executing it in its entirety. Löb's theorem prevents M from "trusting" an agent in its environment, even granting that M has perfect knowledge that the agent reasons correctly within S.

Reflective trust. Alternatively, there can be situations in which it might be relatively easy for M to derive the statement $\mathbb{D}_S[\langle A_0 \rangle]$, and even to produce an algorithm which derivably generates a formal derivation of A_0, but unfeasably difficult for M to actually execute that algorithm. The formal derivation of A_0 might be astronomically long, for example, even while the derivation that it can be constructed is quite short. Again, M finds itself in a situation where it "knows" that it can derive A_0, but is unable to act on this knowledge without first performing some tedious or even unfeasable computation whose result is known in advance. Evidently M cannot even trust itself.

The problems of naturalistic and reflective trust are both straightforward expressions of M's inability to affirm the soundness scheme, and both of them could be easily handled by modifying its licensing criteria. All we have to do is to program M so that whenever it accepts a sentence A_0 as a license to perform the action α_0, it also accepts $\mathbb{D}_S[\langle A_0 \rangle]$ as a license to perform α_0. This would allow it to act on the knowledge that A_0 is derivable — knowledge that it might obtain on its own or from an outside source — without actually possessing a derivation of A_0. Next I present two additional difficulties which cannot be handled in this way.

Reflectively coherent trust. Suppose the actionable condition has the form $(\forall x)A_0[x]$. It could be the case that M is able to derive each instance $A_0[\hat{x}]$, yet not able to derive the quantified statement $(\forall x)A_0[x]$.

There is nothing worrisome about this possibility in itself. It could easily happen that each instance $A_0[\hat{x}]$ is verifiable by a finite computation, yet there is no uniform reason why $A_0[x]$ is true for every x. (Think of statements like "the number $2x + 4$ is a sum of two primes".) But suppose in addition that M knows that for each x it can derive $A_0[\hat{x}]$. That is, suppose M has derived the sentence

$$(\forall x)\mathbb{D}_S[\langle A_0[\hat{x}] \rangle].$$

Without knowing that S is sound, M cannot go on to infer the condition $(\forall x)A_0[x]$ which allows it to act.

One might hope to rig the formal system S so as to evade this problem, by ensuring that whenever $(\forall x)\mathbb{D}_S[\langle A[\hat{x}]\rangle]$ is derivable in S, it is also the case that $(\forall x)A[x]$ is derivable in S. However, reflective coherence in this sense is impossible. As we saw in the proof of Theorem 5.2, PA^A derives a sentence which affirms that for every number x the system PA^A derives that x is not the Gödel number of a derivation in PA^A of $\bot$. The same reasoning shows that this conclusion applies to any finitely decidable extension of PA. If S were such a system then the condition I just suggested would entail that Con(S) is derivable in S, and hence that S is inconsistent.

However, it still seems reasonable to expect M to be able to accept a derivation of $(\forall x)\mathbb{D}_S[\langle A[\hat{x}]\rangle]$ as licensing the same actions that are licensed by a derivation of $(\forall x)A[x]$. The point here is that agreeing to accept $\mathbb{D}_S[\langle(\forall x)A[x]\rangle]$ as a licensing condition, as proposed above, does not accomplish this. We should also note that the issue is no longer about feasability: it may simply be impossible for M to convince itself that $(\forall x)A[x]$ holds, despite knowing $(\forall x)\mathbb{D}_S[\langle A[\hat{x}]\rangle]$.

Disjunctive trust. Finally, suppose the actionable condition is the sentence A_0 and we follow the suggestion made above to program M to also accept $\mathbb{D}_S[\langle A_0\rangle]$, $\mathbb{D}_S[\langle\mathbb{D}_S[\langle A_0\rangle]\rangle]$, etc., as licensing the same action. This does not accomodate the possibility that M might derive the sentence $A_0 \vee \mathbb{D}_S[\langle A_0\rangle]$.[6] Intuitively, the preceding sentence tells us that either A_0 is true, which licenses action, or else A_0 is derivable and therefore true, which also licenses action. So $A_0 \vee \mathbb{D}_S[\langle A_0\rangle]$ "should" license action, but it does not. We cannot straightforwardly infer either A_0 or $\mathbb{D}_S^k[\langle A_0\rangle]$ for any k.

A natural idea is to augment S with a truth predicate. This would enable us to formalize the reasoning we just used which allowed us to infer the truth of A_0 from the derivability of $A_0 \vee \mathbb{D}_S[\langle A_0\rangle]$. The problem is that self-applicative truth predicates are inconsistent, whereas a non self-applicative truth predicate would only apply to reasoning carried out in the original system, not the augmented system. Thus M could still find itself in a situation where it has derived that it can derive A_0, but is unable to infer A_0 because the anticipated derivation of A_0 makes use of the truth predicate. A partially self-applicative truth predicate a la Kripke would not do any better; by Löb's theorem it is simply impossible to consistently augment S in any way that would enable us to generally infer A from $\mathbb{D}_{S+}[\langle A\rangle]$, where $\mathbb{D}_{S+}$ refers to derivability in the augmented system S^+.

As we have seen, the problems of reflective coherence and disjunctive trust are not resolved by broadening the licensing criteria so that $\mathbb{D}_S[\langle A \rangle]$ licenses any action that A licenses. There can still be derivable assertions which ought to license actions but do not. Another possibility is to simply ignore the derivability predicate for licensing purposes; that is, program M so that whenever A licenses α, so does any sentence which reduces to A when all derivability predicates are removed. But this idea is a non-starter because derivably true sentences can become false when derivability predicates are removed. For instance, we can derive in PA that

$$\mathbb{D}_{PA}[\langle \text{Con}(PA) \rangle] \to \mathbb{D}_{PA}[\langle \bot \rangle]$$

(if PA derives its own consistency, then it is inconsistent) but after removing the derivability predicates this becomes

$$\text{Con}(PA) \to \bot,$$

i.e., $\neg\text{Con}(PA)$.

What we need is a principled method of broadening some initially given licensing criteria that goes beyond merely accepting $\mathbb{D}[\langle A \rangle]$ in place of A. This can be done using assertibility. First, we require agents to reason within systems that have an assertibility predicate satisfying the axioms for $\mathbb{A}$ and for which we are able to prove a version of Theorem 5.2. This addresses their inability to affirm the consistency and soundness of their own reasoning. We also impose the following licensing rule.

Assertibility rule: Whenever a sentence A licenses some action α, the sentence $\mathbb{A}[\langle A \rangle]$ also licenses α.

This rule reflects the intuition that $\mathbb{A}[\langle A \rangle]$ is just as good as A; it could also be seen as a version of the release rule discussed in Section 4.5.

The problems described above are now easily resolved. By assertible soundness, whenever $\mathbb{D}_S[\langle A \rangle]$ can be derived, so can $\mathbb{A}[\langle A \rangle]$. Therefore, according to the assertibility rule, knowing that A is derivable is always just as actionable as knowing A. This handles naturalistic and reflective trust. For reflectively coherent trust, observe that

$$(\forall x)\mathbb{D}_S[\langle A[\hat{x}] \rangle] \quad \rightsquigarrow \quad (\forall x)\mathbb{A}[\langle A[\hat{x}] \rangle] \quad \rightsquigarrow \quad \mathbb{A}[\langle (\forall x)A[x] \rangle],$$

derivably in S, where we write $\ulcorner A \rightsquigarrow B \rightsquigarrow C \urcorner$ for $\ulcorner A \to B$ and $B \to C \urcorner$. So $(\forall x)\mathbb{D}_S[\langle A[\hat{x}] \rangle]$ is just as actionable as $(\forall x)A[x]$. Finally, disjunctive trust is handled by the inference

$$A \vee \mathbb{D}_S[\langle A \rangle] \quad \rightsquigarrow \quad A \vee \mathbb{A}[\langle A \rangle] \quad \rightsquigarrow \quad \mathbb{A}[\langle A \rangle],$$

which shows that $A \vee \mathbb{D}_S[\langle A \rangle]$ is just as actionable as A.

Do we get too much? For instance, is $\mathbb{D}_{PA}[\langle \text{Con}(\text{PA}) \rangle] \to \mathbb{D}_{PA}[\langle \perp \rangle]$ just as actionable as $\text{Con}(\text{PA}) \to \perp$? No, because the implication $\mathbb{D}[\langle A \rangle] \to \mathbb{A}[\langle A \rangle]$ only goes in one direction, and besides, we cannot bring an implication inside the assertibility predicate. So there is no way to remove the derivability predicates in this case.

I do not mean to imply that rational agents should be required to work in the language of first order arithmetic. That is merely a convenient vehicle for the assertible soundness and consistency results proven in Theorem 5.2.[7]

5.4   Weak interpretation

Increasing formulas. Weak interpretation and relative consistency. The sequent calculus. A game on formulas. A winning strategy for Defender. Weak interpretation of increasing axioms.

A sentence of the form $L \leftrightarrow \neg L$ is tautologically false; any formal system which included it as an axiom would have to be inconsistent. In contrast, $L \leftrightarrow \neg \mathbb{A}(L)$ characterizes the assertible liar sentence, and we have seen that such things can be reasoned about consistently. Thus, inserting assertibility operators into the axioms of an inconsistent theory can sometimes make it consistent. The aim of this section is to prove a result of the opposite type which shows that axioms of a certain form are in a limited sense indifferent to the insertion of assertibility operators. The motivation is to explore the extent to which we may be justified in ignoring the difference between A and $\mathbb{A}(A)$.

Say that a formula is *increasing* if no implication appears in the premise of any other implication. Note that since we take $\neg A$ to be an abbreviation of $A \to \perp$, this also means that an increasing formula cannot position a negation within the premise of any implication, nor can it contain the negation of any formula containing an implication. The techniques developed below are only able to handle increasing formulas.

Suppose S_1 and S_2 are formal systems such that the language of S_2 is the language of S_1 augmented with the concept symbol $\mathbb{A}$, and assume the axioms for $\mathbb{A}$ are included in S_2. Then let us say that S_2 *weakly interprets* S_1 if assertibility operators can be inserted in any increasing sentence derivable in S_1 in such a way that the resulting sentence is derivable in S_2. Another way to say this is that when the assertibility operators are removed from all sentences derivable in S_2 the resulting set contains every increasing sentence derivable in S_1. Roughly speaking, S_2 weakly interprets S_1 if S_2 interprets S_1 modulo ignoring the difference between A and $\mathbb{A}(A)$ and restricting our attention to increasing sentences.

We will assume the systems do not quantify over sentences. Only in such settings does it even make sense to talk about "removing" appearances of $\mathbb{A}$. Thus, we are making the alternate choice mentioned in Section 4.5 of treating $\mathbb{A}$ as a logical operator to be interpreted as "it is assertible that". The appropriate axioms for $\mathbb{A}$ here are the schematic laws (C_{lo}), $(\wedge_{lo})$, $(\rightarrow_{lo})$, and $(\forall_{lo})$ given in Section 4.5. Also, we will consider A to be a subformula of $\mathbb{A}(A)$ (and so we do not consider $\mathbb{A}(A)$ to be atomic). A specific example of a formal system in which $\mathbb{A}$ appears as a logical operator will be presented in the next section.

We have the following trivial result:

Proposition 5.5. *If* S_2 *weakly interprets* S_1 *then the complete consistency of* S_2 *implies the consistency of* S_1.

This is just because if S_1 were inconsistent then $\perp$ would be derivable in S_1, and weak interpretability would then imply that $\mathbb{A}^k(\perp)$ must be derivable in S_2 for some k. So if S_1 were not consistent then S_2 would not be completely consistent.

Before continuing we need to introduce a new system of deductive reasoning. Hilbert systems lack theoretical simplicity, while natural deduction can be awkward to work with because of the possible nested structure of premises. This suggests that we modify natural deduction so as to explicitly list all current assumptions at each step. But after we do this it becomes natural to treat the assumptions and the consequence in a more symmetric fashion. Gentzen's *sequent calculus* achieves this symmetry by replacing the elimination rules for the conclusion with introduction rules for the assumptions. Remarkably, the resulting system fully captures formal derivability: the sequent calculus is equivalent both to natural deduction and to Hilbert systems in the sense that all three systems derive the same sentences.[8] However, the nonexistence of elimination rules gives this system a kind of transparency that will be very useful.

The basic element of a sequential deduction is a *sequent* $\Gamma \Rightarrow C$ consisting of a finite multiset of formulas Γ, the *premise*, and a single formula C, the *conclusion*. "Multiset" means that Γ is an unordered collection in which formulas may appear more than once. If A is a formula and Γ is a multiset then A, Γ is the same multiset with one additional copy of A. The sequents $A \Rightarrow A$ and $\perp \Rightarrow A$ with A atomic are *trivial*.

A *sequent derivation* is a finite sequence of sequents each of which either is trivial, or is one of the previous sequents in the derivation with a formula added to the premise or with a duplicate formula in the premise deleted, or can be derived from previous sequents in the derivation by one of the following rules:

Premise introduction rules

(P∧) Given either $A, \Gamma \Rightarrow C$ or $B, \Gamma \Rightarrow C$, derive $A \wedge B, \Gamma \Rightarrow C$.

(P∨) Given both $A, \Gamma \Rightarrow C$ and $B, \Gamma \Rightarrow C$, derive $A \vee B, \Gamma \Rightarrow C$.

(P→) Given both $\Gamma \Rightarrow A$ and $B, \Gamma \Rightarrow C$, derive $A \rightarrow B, \Gamma \Rightarrow C$.

(P∀) Given $A[t], \Gamma \Rightarrow C$, derive $(\forall x)A[x], \Gamma \Rightarrow C$.

(P∃) Given $A[x], \Gamma \Rightarrow C$ such that x does not appear freely in any other formula in the sequent, derive $(\exists x)A[x], \Gamma \Rightarrow C$.

Conclusion introduction rules

(C∧) Given both $\Gamma \Rightarrow A$ and $\Gamma \Rightarrow B$, derive $\Gamma \Rightarrow A \wedge B$.

(C∨) Given either $\Gamma \Rightarrow A$ or $\Gamma \Rightarrow B$, derive $\Gamma \Rightarrow A \vee B$.

(C→) Given $A, \Gamma \Rightarrow B$, derive $\Gamma \Rightarrow A \rightarrow B$.

(C∀) Given $\Gamma \Rightarrow A[x]$ such that x does not appear freely in any other formula in the sequent, derive $\Gamma \Rightarrow (\forall x)A[x]$.

(C∃) Given $\Gamma \Rightarrow A[t]$, derive $\Gamma \Rightarrow (\exists x)A[x]$.

As with natural deduction, we require all term substitutions to be free. To avoid confusion I will not refer the rules themselves as having premises and conclusions; rather, I will speak of *incoming* sequents (such as $A, \Gamma \Rightarrow C$ in (P∧)) and *outgoing* sequents (such as $A \wedge B, \Gamma \Rightarrow C$ in (P∧)).

A derivation of a formula A from a set of sentences X is a sequent derivation such the conclusion of the final sequent is A and every formula in the premise of the final sequent belongs to X. I have set up the sequent calculus in a form suitable for intuitionistic reasoning.

I will now describe a procedure for inserting assertibility operators into formulas that lack them and prove that if the initially given formula has a certain form and is derivable in intuitionistic logic then we can ensure that the formula generated by our procedure will be derivable in intuitionistic logic augmented by the axiom schemes (C_{lo}), $(\wedge_{lo})$, $(\rightarrow_{lo})$, and $(\forall_{lo})$. This result can be used to establish weak interpretability results, as I will show in Theorem 5.9 below.

Our procedure can be described as a game between two players, Attacker and Defender, on a formula A_0 that initially lacks assertibility operators.

We think of Attacker as seeking to strengthen the formula and Defender as seeking to weaken it. The way the game is played is defined inductively on the complexity of A_0. The basic move involves prefixing a subformula A with $\mathbb{A}^k$ for some value of $k \geq 0$, taking this to mean leaving A unaltered if $k = 0$. For brevity I will simply call this action "prefixing A". If A_0 is atomic then the game consists of a single move in which Defender prefixes A_0; Attacker does not have a turn. The game is played on formulas of the form $A \wedge B$ by independently playing the game on A and B and conjoining the results, then allowing Defender to prefix this conjunction. It is played on $A \vee B$ by independently playing on A and B and disjoining the results, then allowing Defender to prefix this disjunction. It is played on $(\exists x)A$ and $(\forall x)A$ by playing on A, adding the relevant quantifier, then allowing Defender to prefix the result. Finally, it is played on $A \rightarrow B$ by first having the players switch roles and play on A, producing a formula A', then revert to their original roles and play on B, producing a formula B', and then allowing Defender to prefix $A' \rightarrow B'$. Observe that if the original formula contains no implications then Attacker never gets a turn.

We require a simple lemma. If A is a formula possibly involving $\mathbb{A}$, let $\breve{A}$ denote the formula obtained by deleting all instances of $\mathbb{A}$. Also, recall that $\ulcorner A \rightsquigarrow B \rightsquigarrow C \urcorner$ abbreviates $\ulcorner A \rightarrow B$ and $B \rightarrow C \urcorner$.

Lemma 5.6. *Let A be a formula in which $\rightarrow$ does not appear. Then*

$$\breve{A} \quad \rightsquigarrow \quad A \quad \rightsquigarrow \quad \mathbb{A}^j(\breve{A})$$

for some $j \geq 0$.

Proof. Both implications are proven by induction on the complexity of A. The first follows trivially from repeated application of capture. The second is trivial if A is atomic because then $\breve{A} = A$. Next, suppose A has the form $A_1 \wedge A_2$. If inductively $A_1 \rightarrow \mathbb{A}^{j_1}(\breve{A}_1)$ and $A_2 \rightarrow \mathbb{A}^{j_2}(\breve{A}_2)$ then

$$A_1 \wedge A_2 \quad \rightsquigarrow \quad \mathbb{A}^j(\breve{A}_1) \wedge \mathbb{A}^j(\breve{A}_2) \quad \rightsquigarrow \quad \mathbb{A}^j(\breve{A}_1 \wedge \breve{A}_2)$$

where $j = \max(j_1, j_2)$, by multiple applications of capture in the first step and of the $(\wedge)$ and $(\rightarrow)$ laws in the second step. The other logical symbols are treated similarly, applying Corollary 4.3 (b) and (e) and the $(\forall)$ law. Finally, the case where A has the form $\mathbb{A}(B)$ is easy because then $\breve{A} = \breve{B}$, so if $B \rightarrow \mathbb{A}^j(\breve{B})$ then $A \rightarrow \mathbb{A}^{j+1}(\breve{A})$. This completes the proof. $\square$

If the initially given formula is derivable in intuitionistic logic, we consider Defender to win provided the formula generated by the game is derivable in intuitionistic logic augmented by the axioms for $\mathbb{A}$. We will say that a player has a *winning strategy* if that player is able to win against any sequence of moves by the opponent.

Theorem 5.7. *Defender has a winning strategy on any formula of the form* $A_0 \to B_0$, *with* A_0 *and* B_0 *increasing, that is intuitionistically derivable.*

Proof. A strategy for either player is given by specifying, for each of his prefixing moves, the value of k to be played as a function of the values chosen by his opponent on all of the opponent's earlier moves.

We can order strategies by saying that $S \leq S'$ if, at every play, for each choice of earlier moves by the opponent, the value of k prescribed by S' is greater than or equal to the value prescribed by S. I claim that if Attacker plays the same moves against two of Defender's strategies, S and S', such that $S \leq S'$, then the formula generated by the first game will imply the formula generated by the second game, and if Defender plays the same moves against two of Attacker's strategies, $\mathcal{T}$ and $\mathcal{T}'$, such that $\mathcal{T} \leq \mathcal{T}'$, then the formula generated by the first game will be implied by the formula generated by the second game. This is shown by a straightfoward induction on the complexity of the formula on which the game is played, taking both claims for all simpler formulas as the induction hypothesis. It follows that any strategy greater than a winning strategy also wins.

Next, we may extend the definition of the game so that it can be played on sequents. Here the game is played by first having Attacker and Defender switch roles and play the game independently on each of the premise formulas, then revert to their original roles and play the game on the conclusion. Defender wins if it is derivable in intuitionistic logic plus the axioms for assertibility that the conjunction of the resulting premise formulas implies the resulting conclusion.

Now if $A_0 \to B_0$ is intuitionistically derivable then it has a sequent derivation whose penultimate sequent is $A_0 \Rightarrow B_0$, followed by the final sequent $\emptyset \Rightarrow (A_0 \to B_0)$. So it will suffice to show that Defender has a winning strategy on the sequent $A_0 \Rightarrow B_0$. Moreover, it is a general fact about sequent derivations that we can assume all formulas previously appearing in this derivation are subformulas of A_0 and B_0. If A_0 and B_0 are increasing, this means that all formulas in the derivation which precede $A_0 \Rightarrow B_0$ are increasing.

The proof is completed by checking that Defender has a winning strategy on any trivial sequent, if Defender has a winning strategy on a given sequent then he has a winning strategy on the same sequent with an additional formula in the premise or with a duplicate formula in the premise deleted, and if Defender has a winning strategy on the incoming sequents of a rule then he has a winning strategy on the outgoing sequent of that rule — all of this assuming that all formulas that appear are increasing.

Given a trivial sequent $A \Rightarrow A$, on the first move Attacker will prefix the A in the premise and on the second move Defender will prefix the A in the conclusion. Defender wins by using the same value of k used by Attacker. The same idea works on trivial sequents of the form $\bot \Rightarrow A$.

If a formula is added to the premise of a sequent on which Defender has a winning strategy, then Defender can play any strategy on it while retaining his previous strategy on the rest of the sequent and still win. If a duplicate formula A in the premise is deleted, Defender should play on the remaining A the max of the two strategies he played on the previous two copies of A, and retain his previous strategy on the rest of the sequent. It is easy to see that here too Defender still wins.

All of the conclusion introduction rules and all of the premise introduction rules besides $(P\rightarrow)$ are straightforwardly handled in a similar way, modifying the strategies used in the incoming sequents either not at all, or at worst, taking the max of two strategies on formulas which appear in more than one incoming sequent. For $(P\rightarrow)$, observe first that since $A \rightarrow B$ is increasing A cannot contain any implications. We know inductively that Defender has a winning strategy on $\Gamma \Rightarrow A$, that is, he can ensure that the conjunction of the formulas in Γ' implies A', where $\Gamma' \Rightarrow A'$ is the result of playing on the sequent $\Gamma \Rightarrow A$. Thus by Lemma 5.6 Defender can ensure that the conjunction of the formulas in Γ' implies $\mathbb{A}^j(A)$ for some j. (However, the value of j may depend on Attacker's choice of moves in Γ.) We also know inductively that Defender can win on the sequent $B, \Gamma \Rightarrow C$. To play on the sequent $A \rightarrow B, \Gamma \Rightarrow C$, Defender may use a null strategy on A, the strategy previously used on B, and the maximum of the strategies used on the two copies of Γ. If the previous strategy used on C is retained, then this leads to a premise $A \rightarrow B', \Gamma'$ with the property that the conjunction of the formulas in Γ' implies $\mathbb{A}^j(A)$ for some A, and this conjunction together with B' implies C'. But $\mathbb{A}^j(A) \wedge (A \rightarrow B') \rightarrow \mathbb{A}^j(B')$, and $\mathbb{A}^j(B')$ plus the conjunction of the formulas in Γ' implies $\mathbb{A}^j(C')$. So Defender needs only to alter the previous strategy used on C by increasing the power of the prefix by j in the final step. This shows that, assuming the induction hypothesis, Defender can ensure that the result of playing on the sequent $A \rightarrow B, \Gamma \Rightarrow C$ is a sequent whose conclusion follows from the conjunction of the premises, as desired. □

In order to apply this theorem we need one more easy lemma. Recall that $\check{A}$ is the formula obtained by deleting all instances of $\mathbb{A}$ from A.

Lemma 5.8. *Let A be an increasing formula. If the game is played on $\check{A}$ then Defender can ensure that the resulting formula A' satisfies $A \to A'$.*

Proof. The proof goes by induction on the complexity of $\check{A}$. If $\check{A}$ is atomic then $A = \mathbb{A}^j(\check{A})$ for some j and Defender can prefix $\check{A}$ with $\mathbb{A}^j$ to obtain $A' = A$. In general, if A has the form $\mathbb{A}(B)$ then Defender plays as on B but prefixes with one additional $\mathbb{A}$ in the final move; this works inductively because $B \to B'$ implies $\mathbb{A}(B) \to \mathbb{A}(B')$, by capture and ($\to$). If A has the form $A_1 \wedge A_2$ then Defender plays as before on A_1 and A_2 and adds no final prefix; this works because $A_1 \to A_1'$ and $A_2 \to A_2'$ imply $A_1 \wedge A_2 \to A_1' \wedge A_2'$. The cases $A_1 \vee A_2$, $(\forall x)B$, and $(\exists x)B$ are handled similarly.

The only interesting case is where A has the form $A_1 \to A_2$. Since A is increasing, $\to$ cannot appear in A_1. Thus, when playing on $A_1 \to A_2$ we first play on A_1 and have from Lemma 5.6 that the resulting formula A_1' (whose form is determined entirely by Attacker) satisfies $A_1' \rightsquigarrow \mathbb{A}^j(\check{A}_1) \rightsquigarrow \mathbb{A}^j(A_1)$ for some j. Since the overall induction in this proof is on the complexity of $\check{A}$, the induction hypothesis provides that Defender can then employ a strategy on $\check{A}_2$ which ensures that $\mathbb{A}^j(A_2) \to A_2'$. This yields that $A_1 \to A_2$ implies

$$A_1' \rightsquigarrow \mathbb{A}^j(A_1) \rightsquigarrow \mathbb{A}^j(A_2) \rightsquigarrow A_2'.$$

Thus we obtain $A \to A'$ when A is $A_1 \to A_2$ (with no final prefixing move on $\check{A}_1 \to \check{A}_2$ needed). $\qquad\qquad\qquad\qquad\qquad\qquad\qquad\qquad\square$

Theorem 5.9. *Suppose the nonlogical axioms of S_2 are increasing and the nonlogical axioms of S_1 are those of S_2 with all assertibility operators deleted. Then S_2 weakly interprets S_1.*

Proof. Let B be any increasing sentence derivable in S_1. We must find a way to insert assertibility operators into B so that it becomes derivable in S_2. Since B is derivable in S_1, it is a logical consequence of finitely many nonlogical axioms $\check{A}_i$ of S_1 corresponding to axioms A_i of S_2; writing their conjunction as $\check{A} = \bigwedge \check{A}_i$, we therefore have that $\check{A} \to B$ is intuitionistically derivable. The problem is then to insert assertibility operators into $\check{A}$ and B, yielding new formulas A' and B', in such a way that $A' \to B'$ is derivable in intuitionistic logic augmented by the axioms for $\mathbb{A}$, and such that each conjunct A'_i of A' is implied by the corresponding nonlogical axiom A_i of S_2. It will then follow that B' is derivable in S_2, as desired.

We achieve this result by playing the game described above on the formula $\check{A} \to B$. We know by Theorem 5.7 that Defender can ensure the resulting formula $A' \to B'$ is derivable in intuitionistic logic augmented by the axioms for $\mathbb{A}$, so all we need to do is to prescribe a strategy for Attacker which ensures that each conjunct A'_i of A' is implied by the corresponding axiom A_i.

The game is played on $A \to B$ by first switching the players' roles and playing on A. So we reduce to a problem about finding a strategy for Defender when the game is played on $\check{A}$, and this amounts to giving a strategy for Defender on each conjunct $\check{A}_i$ in a way that ensures $A_i \to A'_i$. This strategy is provided by Lemma 5.8. □

5.5 Conceptual assertibility

The system CC. *It formalizes the Russell concept. It is completely consistent. Second-order Peano arithmetic.* $\mathrm{Comp}(\mathrm{PF}_\mathcal{T}) + \mathrm{D}$.

The basic problem with second-order logic is that it makes no sense to quantify over schematic variables, and the solution to this problem is to use an assertibility predicate to give assertoric force to ordinary variables. Thus if C is an ordinary variable that ranges over concepts and x is an ordinary variable that ranges over objects, then $C[[x]]$ is a term which stands for an atomic proposition, and $\mathbb{A}(C[[x]])$ is an atomic sentence which expresses that x falls under C. This is according to the global notion of falling under that I described in Section 4.6. Thus $\mathbb{A}(C[[x]])$ affirms that the sentence which ascribes C to x is assertible.

I will now present a simple system for second-order reasoning about concepts. The language has an infinite list of variables x, y, z, ..., and the atomic formulas have the form $\bot$, $x \, \varepsilon \, y$, and $x = y$, with any variables in place of x and y. There are no terms besides the individual variables. General formulas are built up from the atomic formulas by the following rules: if A and B are formulas then so are $(A \wedge B)$, $(A \vee B)$, $(A \to B)$, $(\forall x)A$, $(\exists x)A$, and $(\mathbb{A}A)$.

The variables are to be thought of as ranging over concepts. Thus, there are no objects in this system besides concepts. The notation "$x \, \varepsilon \, y$" is supposed to express "the sentence which ascribes y to x is assertible", and the notation $\ulcorner \mathbb{A}A \urcorner$ is supposed to express $\ulcorner$it is assertible that $A\urcorner$. We may consider $\perp$ to express the statement that every sentence is assertible.

The language is identical to the language of set theory augmented by an assertibility predicate, reflecting the fact that we talk about objects falling under concepts in much the same way that we talk about objects belonging to sets. I use the variant "ε" notation to emphasize that $x \, \varepsilon \, y$ indicates a form of falling under, not membership.

Observe that $\mathbb{A}$ functions as a logical operator, not a predicate, and it applies to arbitrary formulas. Thus, e.g., $\mathbb{A}(x = y)$ is not a sentence, while $(\forall x)(\exists y)\mathbb{A}(x = y)$ is (parenthesized for readability).

The axioms consist of the universal closures of the following:

(C) $A \to \mathbb{A}A$
($\wedge$) $\mathbb{A}A \wedge \mathbb{A}B \to \mathbb{A}(A \wedge B)$
($\to$) $\mathbb{A}A \wedge \mathbb{A}(A \to B) \to \mathbb{A}B$
(=) $x = y \leftrightarrow (\forall u)(u \, \varepsilon \, x \leftrightarrow u \, \varepsilon \, y)$
(Comp) $(\exists y)(\forall x)(x \, \varepsilon \, y \leftrightarrow \mathbb{A}A)$

where A and B are arbitrary formulas, except in (Comp), where A can be any formula in which y does not appear freely. Thus (=) is a single axiom and the rest are schemes.

The schemes (C), ($\wedge$), and ($\to$) are the familiar axioms for assertibility as a logical operator. The axiom (=) is a version of extensionality which expresses a slightly weaker notion of equality than that described in Section 4.6. (In order for C and D to be considered equal, we do not have to be able to affirm the assertibility of a single sentence that expresses an equivalence between $C[[x]]$ and $D[[x]]$ for all x, but merely to be able to affirm that for every x there is an equivalence between the assertibility of $C[[x]]$ and the assertibility of $D[[x]]$. This is the more natural condition given the constraints of the language we are working with.)

(Comp) is a comprehension scheme which formalizes the idea that every formula $A[x]$ with one free variable x expresses a concept. Thus, for example, if $A[x]$ is the formula "x is white" then A expresses the concept of whiteness, and for any x the sentence $A[\hat{x}]$ ascribes whiteness to x. So we can say that $A[\hat{x}]$ is assertible if and only if the ascription of whiteness to x is assertible. For general A the statement would be that there is a concept y such that for all x the ascription of y to x is assertible if and only if $A[x]$ is assertible. This should explain the motivation for (Comp) when A contains no free variables besides x. If A contains other free variables, then we can consider those other variables as parameters and observe that A will express a concept for any choice of those parameters. So (Comp) remains valid.

We adopt intuitionistic logic and include the usual axioms for equality (the first three Peano axioms: reflexivity, symmetry, and transitivity, plus the intersubstitutability axiom $x = y \to (x \in z \leftrightarrow y \in z)$). I will call the resulting formal system *Constructive Concepts (CC)*. It is notable in having a full comprehension scheme. Thus, for instance, we can exhibit a version of the Russell concept via the formula $\neg(x\,\varepsilon\,x)$. According to (Comp), there exists a concept R such that

$$x\,\varepsilon\,R \quad \leftrightarrow \quad \mathbb{A}(x \not\varepsilon x).$$

This concept skirts paradoxicality in the same way that the assertible liar sentence does. Assuming $R \not\varepsilon R$ yields $\mathbb{A}(R \not\varepsilon R)$ by capture, which entails $R\,\varepsilon\,R$ by the characterization of R. This shows that $R \not\varepsilon R$ is contradictory, so we conclude $\neg(R \not\varepsilon R)$. On the other hand, assuming $R\,\varepsilon\,R$ immediately yields $\mathbb{A}(R \not\varepsilon R)$, but since $R\,\varepsilon\,R$ also implies $\mathbb{A}(R\,\varepsilon\,R)$, we infer $\mathbb{A}\bot$. So we have $R\,\varepsilon\,R \to \mathbb{A}\bot$. The sentence $R \not\varepsilon R$ is false and the sentence $R\,\varepsilon\,R$ is weakly false.

Thus, we can reason in CC about apparently paradoxical concepts and reach substantive conclusions. But no contradiction can be derived, by the following result. Its proof follows a familiar pattern.

Theorem 5.10. CC *is completely consistent.*

Proof. Fix a distinguished variable r. We begin by adding countably many constants to the language of CC. Let $\mathbb{L}$ be the smallest language which contains the language of CC and which contains, for every formula A of $\mathbb{L}$ in which no variable other than r appears freely, a constant symbol c_A. Observe that $\mathbb{L}$ is countable.

We define a sequence of sets of sentences $(\mathcal{F}_n)$. We include $\bot$ in $\mathcal{F}_n$ for all n; we place $c_B\,\varepsilon\,c_A$ in $\mathcal{F}_n$ if $A[c_B]$ belongs to $\mathcal{F}_{n-1}$; we place $c_A = c_{A'}$ in $\mathcal{F}_n$ if for some c_B, one but not both of $A[c_B]$ and $A'[c_B]$ belongs to $\mathcal{F}_{n-1}$; and we place $\mathbb{A}A$ in $\mathcal{F}_n$ if A belongs to $\mathcal{F}_{n-1}$. No sentence of the form $c_B\,\varepsilon\,c_A$, $c_A = c_{A'}$, or $\mathbb{A}A$ is placed in $\mathcal{F}_0$. (Recall that the constants c_A are only defined for formulas A in which no variable other than r appears freely. So expressions like $A[c_B]$ are unambiguous.) For more complex sentences we apply the usual rules. $A \wedge B$ belongs to $\mathcal{F}_n$ if either A or B belongs to $\mathcal{F}_n$. $A \vee B$ belongs to $\mathcal{F}_n$ if both A and B belong to $\mathcal{F}_n$. $(\forall x)A$ belongs to $\mathcal{F}_n$ if $A[c_B]$ belongs to $\mathcal{F}_n$ for some constant c_B, and $(\exists x)A$ belongs to $\mathcal{F}_n$ if $A[c_B]$ belongs to $\mathcal{F}_n$ for every constant c_B. (Note here that if $(\forall x)A$ is a sentence then A can contain no free variables other than x, so again the expression $A[c_B]$ is unambiguous.) Finally, $A \to B$ belongs to $\mathcal{F}_n$ if there exists $k \leq n$ such that B belongs to $\mathcal{F}_k$ but A does not belong to $\mathcal{F}_k$.

By the usual argument the sequence $(\mathcal{F}_n)$ is increasing, and it is clear that $\mathbb{A}^k \bot$ belongs to $\mathcal{F} = \bigcup \mathcal{F}_n$ for all k. The proof is completed by checking that no axiom of CC belongs to $\mathcal{F}$ and the complement of $\mathcal{F}$ is stable under universal modus ponens. This is mostly similar to the corresponding arguments in the proofs of complete consistency of PAR and PA$^{\mathbb{A}}$. Verifying that no instance of comprehension belongs to $\mathcal{F}$ is straightforward; this reduces to checking that $(\forall r)(r \in c_A \leftrightarrow \mathbb{A}A)$ is not in $\mathcal{F}$ for any formula A with no free variables besides r. The point is that if either $c_B \in c_A$ or $\mathbb{A}(A[c_B])$ fails then the other fails at the same stage, namely one stage after $A[c_B]$ fails. Extensionality is also easy because our criterion for placing $c_A = c_{A'}$ in $\mathcal{F}_n$ is specifically designed to test for failure of the condition $c_B \, \varepsilon \, c_A \leftrightarrow c_B \, \varepsilon \, c_{A'}$. Intersubstitutability is more interesting because there is a concern that $c_A \, \varepsilon \, c_B \leftrightarrow c_{A'} \, \varepsilon \, c_B$ could fail for some B before $c_A = c_{A'}$ fails, if the latter fails at all. But in fact this cannot happen because if n is the smallest value for which there exists B such that $c_A \, \varepsilon \, c_B$ belongs to $\mathcal{F}_n$ but $c_{A'} \, \varepsilon \, c_B$ does not, or vice versa, then $n-1$ is the smallest value such that $B[c_A]$ belongs to $\mathcal{F}_{n-1}$ but $B[c_{A'}]$ does not, and then there must be some atomic formula which distinguishes c_A from $c_{A'}$ at stage $n-1$ — if no atomic formula distinguishes them then no formula distinguishes them. So either there exists C such that $c_A \, \varepsilon \, c_C$ is in $\mathcal{F}_{n-1}$ but $c_{A'} \, \varepsilon \, c_C$ is not, or vice versa, which contradicts the minimality of n, or else there exists C such that $c_C \, \varepsilon \, c_A$ is in $\mathcal{F}_{n-1}$ but $c_C \, \varepsilon \, c_{A'}$ is not, or vice versa, which forces $c_A = c_{A'}$ to belong to $\mathcal{F}_{n-1}$, or else there exists C such that $c_C = c_A$ is in $\mathcal{F}_{n-1}$ but $c_C = c_{A'}$ is not, or vice versa, which again forces $c_A = c_{A'}$ to belong to $\mathcal{F}_{n-1}$. So we know that c_A and $c_{A'}$ are unequal before we know that they fall under different concepts. $\qquad\square$

There has been a long-standing philosophical debate, going back to Frege and Russell, over the extent to which it is possible to base mathematical reasoning on the pure logic of concepts. But this debate has been quite empty because of a failure to understand how to properly quantify over concepts. In order to give global concept variables assertoric force we must introduce an assertibility operator or predicate.

The system CC should give us some idea of how far we can get mathematically with the pure logic of concepts. The result is disappointing. The simplicity of the consistency proof given in Theorem 5.10 already reveals that CC must be a very weak system. I will now present two positive results which show how (extensions by definitions of) CC can, in the terminology of Section 5.4, weakly interpret more standard formal systems in which the assertibility operator does not appear.

Observe that deleting all assertibility operators in all theorems of CC yields an inconsistency: as we saw earlier, we can derive in CC the existence of a concept R which satisfies both $\neg\neg(R \varepsilon R)$ and $R \varepsilon R \to \mathbb{A}\bot$, and deleting the $\mathbb{A}$ in the second formula produces the contradictory conclusions $\neg\neg(R \varepsilon R)$ and $\neg(R \varepsilon R)$. Nonetheless, no inconsistent theory can be weakly interpreted in CC, by Proposition 5.5.

Enlarge the language of CC by adding a constant symbol 0 which satisfies

$$y \varepsilon 0 \leftrightarrow \mathbb{A}\bot,$$

a unary operation symbol $'$ which satisfies

$$y \varepsilon x' \leftrightarrow \mathbb{A}(y = x),$$

and a constant symbol ω which satisfies

$$y \varepsilon \omega \leftrightarrow \mathbb{A}(\forall z)[(0 \varepsilon z) \wedge (\forall x)(x \varepsilon z \to x' \varepsilon z) \to y \varepsilon z].$$

The following formulas are easily derived in the resulting extension CC$'$ of CC:

(1) $0 \varepsilon \omega$;
(2) $x \varepsilon \omega \to x' \varepsilon \omega$;
(3) $x' = 0 \to \mathbb{A}\bot$;
(4) $x' = y' \to \mathbb{A}(x = y)$;
(5) $(0 \varepsilon z) \wedge (\forall x)(x \varepsilon z \to x' \varepsilon z) \to (\forall x)(x \varepsilon \omega \to \mathbb{A}(x \varepsilon z))$.

These formulas resemble the axioms of second-order Peano arithmetic. This is a version of Peano arithmetic that includes both the usual number variables and variables for sets of numbers, and in which the induction scheme takes the form of the single axiom

$$(0 \in X) \wedge (\forall x)(x \in X \to x' \in X) \to (\forall x)(x \in X).$$

In fact, the axioms of second-order Peano arithmetic are just the universal closures of the formulas (1) — (5) with all appearances of $\mathbb{A}$ deleted. Since the formulas (1) — (4) are increasing, Theorem 5.9 yields the following result.

Theorem 5.11. CC$'$ *weakly interprets intuitionistic second-order Peano arithmetic minus induction.*

(In Theorem 5.9 it is assumed that the system S_2 includes the ($\forall$) scheme, which CC' does not, but careful inspection of the proof of that result reveals that this assumption is not needed here because universal quantifiers do not appear in (1) — (4).)

Since the induction axiom is not increasing it has to be excluded from Theorem 5.11. Thus, although CC' derives a version of full second-order induction, it nonetheless appears to possess only meager number theoretic resources.

The second system we consider, $\mathrm{Comp}(\mathrm{PF}_\mathcal{T}) + \mathrm{D}$, has been shown to be consistent under classical logic.[9] Here we will show that the intuitionistic version of this system is weakly interpretable in an extension by definitions of CC plus the ($\forall$) scheme.

$\mathrm{Comp}(\mathrm{PF}_\mathcal{T}) + \mathrm{D}$ is a "positive" set theory. Its language is the ordinary language of set theory — the language of CC minus the assertibility operator — augmented by terms which are generated in the following way. Any variable is a term; if t_1 and t_2 are terms then $t_1 \, \varepsilon \, t_2$ and $t_1 = t_2$ are positive formulas; if A and B are positive formulas then $A \wedge B$, $A \vee B$, $(\forall x)A$, and $(\exists x)A$ are positive formulas; if A is a positive formula and x is a variable then $\{x : A[x]\}$ is a term whose variables are the free variables of A other than x. The system consists of the comprehension scheme

$$y \, \varepsilon \, \{x : A[x]\} \leftrightarrow A[y],$$

where A ranges over the positive formulas, together with the axiom D which states

$$(\exists x)(\exists y)(x \neq y).$$

Let $CC_\forall$ be CC plus the scheme ($\forall$). This addition could be justified in the same way we justified it for $PA^\mathbb{A}$, after reinterpreting the variables of CC to range not over all concepts but only over the countable family of concepts which can be expressed in a certain extension of $CC_\forall$ which I will describe now.

Recursively add to $CC_\forall$, for every formula A, the term $\{x : \mathbb{A}A[x]\}$ (whose variables are the free variables of A other than x) together with the axiom

$$y \, \varepsilon \, \{x : \mathbb{A}A[x]\} \leftrightarrow \mathbb{A}A[y].$$

I will call the resulting system $CC'_\forall$.

Observe that the axiom $y \, \varepsilon \, \{x : \mathbb{A}A[x]\} \leftrightarrow \mathbb{A}A[y]$ is increasing in the sense of Section 5.4 if A is positive, and the formula $(\exists x)(\exists y)(x = y \rightarrow$

$\mathbb{A}\perp$), which is easily derivable in CC, is also increasing. Since removing all assertibility operators from these formulas recovers the axioms of $\text{Comp}(\text{PF}_{\mathcal{T}}) + \text{D}$, Theorem 5.9 yields the following.

Theorem 5.12. $\text{CC}'_{\forall}$ *weakly interprets intuitionistic* $\text{Comp}(\text{PF}_{\mathcal{T}}) + \text{D}$.

It is interesting to note that the extensionality axiom of CC is not increasing, so that we cannot weakly interpret $\text{Comp}(\text{PF}_{\mathcal{T}}) + \text{D}$ plus extensionality in $\text{CC}'_{\forall}$. The latter theory is in fact inconsistent.[10]

5.6   Second-order logic

A more expressive concept language. Interpretation in $\text{PA}^{\mathbb{A}}$.

In the system CC assertibility is treated as a logical operator, so that $\ulcorner \mathbb{A}A \urcorner$ can be read as $\ulcorner$it is assertible that $A \urcorner$. As a result, we have no ability to manipulate syntax; our only atomic statements are of the form "$C[[x]]$ is assertible". Thus, for example, the formula $x \not\in x$ expresses an assertible Russell concept which says that the sentence which ascribes x to itself is not assertible, but we are unable to express in CC a variant Russell concept which says that the negation of the sentence which ascribes x to itself is assertible.

The advantages of the approach used in CC are its simplicity and the resemblance of its language to the familiar language of set theory. It is also adapted to the results of Section 5.4, which only apply to assertibility systems in which $\mathbb{A}$ is a logical operator. However, it is not difficult to formulate alternative systems for second-order reasoning which treat $\mathbb{A}$ as a concept symbol. I will briefly indicate how this can be done.

First, we build up a repertoire of terms. We still have an infinite list of concept variables x, y, z, ..., but these do not count as terms. The atomic terms consist of $\perp$ and all expressions of the form $x[[y]]$, which is interpreted as a statement that ascribes x to y. In such a term the variable x is *assertoric*, but y is not.

If t_1 and t_2 are terms, then so are $(t_1 \wedge t_2)$, $(t_1 \vee t_2)$, and $(t_1 \rightarrow t_2)$. The assertoric variables in each of these terms are the union of the assertoric variables of t_1 and the assertoric variables of t_2. If t is a term then so is $\dot{\mathbb{A}}[t]$, and this term has no assertoric variables. Finally, if t is a term in which x is not assertoric then $(\dot{\forall}x)t$ and $(\dot{\exists}x)t$ are also terms; they have the same assertoric variables as t. We cannot quantify assertoric variables.

This completes the description of the terms of the language.

We will consider both x and y to be free variables in the term $x[[y]]$, although only x is assertoric. The free variables of $(t_1 \mathbin{\dot{\wedge}} t_2)$, $(t_1 \mathbin{\dot{\vee}} t_2)$, and $(t_1 \mathbin{\dot{\rightarrow}} t_2)$ are the union of the free variables of t_1 and the free variables of t_2, and the free variables of $\dot{\mathbb{A}}[t]$ are the same as the free variables of t. The free variables of $(\dot{\forall}x)t$ and $(\dot{\exists}x)t$ are the free variables of t excluding x.

Next, the atomic formulas of the language are $\perp$ and all expressions of the form $\mathbb{A}[t]$ with t a term. General formulas are built up in the same way as the terms, with the same rule that only unassertoric variables can be quantified. Thus $(\forall x)(\forall y)\mathbb{A}[x[[y]]]$ ("for all x and y, the sentence that ascribes x to y is assertible") and $(\forall x)\mathbb{A}[(\dot{\forall}y)x[[y]]]$ ("for all x, the sentence that ascribes x to y for all y is assertible") are legal sentences, but $\mathbb{A}[(\dot{\forall}x)(\dot{\forall}y)x[[y]]]$ is not, because there is no sentence that ascribes x to y for all x and y. The variant Russell concept is expressed by the formula

$$\mathbb{A}[x[[x]] \mathbin{\dot{\rightarrow}} \dot{\perp}].$$

The schemes (C_{lo}), $(\wedge_{lo})$, and $(\rightarrow_{lo})$ from Section 4.5 would be appropriate. The comprehension scheme could take the form

$$(\exists x)\mathbb{A}[(\dot{\forall}y)(x[[y]] \mathbin{\dot{\leftrightarrow}} t)]$$

with t ranging over all terms in which x does not appear freely and in which y is not assertoric.

A complete consistency proof could then be given along the lines of Theorem 5.10, but this is not necessary because the system described above can be interpreted in $\mathrm{PA}^{\mathbb{A}}$, so complete consistency follows from Theorem 5.3. We can do this by interpreting the variables as ranging over the Gödel numbers of all formulas in the language of $\mathrm{PA}^{\mathbb{A}}$ with one free variable, and interpreting the term-building operations as operations on Gödel numbers.

Chapter 6

Surveyability

6.1 The iterative conception

*Sets versus collections. The iterative conception. How does one "form"
a set? Forming power sets of infinite sets.*

Sets have appeared here and there in previous chapters, but they have
not played a central role. We have mainly been interested in concepts. One
might have expected the two notions to be basically equivalent, such that
for every set X there would be a corresponding concept of belonging to
X, and for every concept C there would be a corresponding set of objects
which fall under C (the "extension" of C). However, we now know that the
second part is wrong: there are concepts which do not determine sets in this
way. Most obviously, the concept of being a set may or may not have an
extension, but if it does, this extension is certainly not a set. Concepts are
more general than sets: they can have a circular character, like the concept
of being a concept, which sets cannot have. Intuitionistic logic must be
used to reason about general concepts, whereas it is basic to our idea of
sets that any given object definitely either does or does not lie in a given
set.

So we need to clarify the relationship between concepts and sets. But
looking to the mathematical literature for guidance about the nature of sets
yields disappointing results. In elementary sources one often encounters the
attitude that sets are familiar objects from everyday life with which we are
already acquainted. According to Halmos's classic text, for example, "A
pack of wolves, a bunch of grapes, or a flock of pigeons are all examples of
sets of things",[1] to which Black responds, "It ought then to make sense,
at least sometimes, to speak of being pursued by a set, or eating a set, or
putting a set to flight".[2]

Black's comment is facetious, but the serious point is that, contrary to Halmos, collective expressions in ordinary language do not really correspond to sets in the mathematical sense. The point was developed by Slater.[3] A term like "deck of cards" behaves as if it references not a set but something called a *mereological sum* — a physical object composed of the individual cards in the same way that a house is made out of bricks. A deck of cards is not a set of cards in the mathematical sense, and the same is true of analogous collective expressions. You can eat a bunch of grapes — not the elements of the bunch, the bunch itself — because it is a physical object, and "the fact that such collections are mereological sums is also shown by the fact that shoals, herds, packs, tribes, and the like, are located and can move about in physical space, just like their members".[4]

In contrast, the word "pair" in the phrase "pair of apples" is a *count noun* which measures the number of apples but does not refer to a separate object. This is most clearly seen by comparing it with a *mass noun* such as "half" in the phrase "half a loaf of bread". For "in 'There is half of a loaf' there is obviously no reference to something other than bread; there is not, in addition, reference to one of a range of mysterious, further objective entities, 'halves', 'quarters', 'parts', etc. There is merely a specification of how much of a loaf there is".[5]

After reading Slater's careful analysis of the various types of collective expressions used in ordinary language and his demonstration that none of them function as mathematical sets, it is hard to avoid the conclusion that the textbook mathematical literature which treats sets as familiar, everyday objects is basically fraudulent. That is not to say that we cannot invent a new type of collective expression which functions as a mathematical set, but the classical paradoxes of naive set theory suggest that there is real work to be done here.

Indeed, the key feature of the standard explanation of the paradoxes is that there is more than one kind of "collection". Not all collections are sets: some, like the collection of all ordinals, are proper classes. This is not a complete solution since one then has the task of explaining why the collection of all classes is not a class, or else why there is no such collection, but it does seem to capture something important about the situation. At least for a platonist, something clearly goes wrong when we consider the collection of all ordinals to be a set that does not go wrong when we consider the collection of all natural numbers to be a set.

Some might say that proper classes are not genuine objects at all, that talk about proper classes is merely a convenient abuse of language that could, if one wanted, always be bypassed in favor of direct reference to sets. Now questions about the legitimacy of abstract objects are a bit murky, but in any case rejecting proper classes as illusory only eliminates the puzzle of having to resolve the notion of a collection into two categories. It does not get to the heart of the difficulty because we still have to understand why some concepts (like being a natural number) determine collections while others (like being an ordinal) do not.

In some of the literature on this subject there is a puzzling tendency to treat the set/proper class distinction as an empirical question, as if the coalescence of some collections, but not others, into sets were simply a natural phenomenon that need not have any special explanation. To the contrary, I feel that we should seek an account of what is meant by the word "set" that is sufficiently informative to give us real insight into why some collections are sets while others are not.

Once we get past the "sets are collections" level of discussion, the most popular explanation of the nature of sets is the *iterative conception*. Just what this conception is, however, is a little unclear. The following selection of commentary should get across the general idea:

> ... sets are just those collections which can be "built up" in stages by the repeated application of certain "set-building" operations ... [6]
>
> ... sets are "formed", "constructed", or "collected" from their elements in a succession of stages... [7]
>
> According to the iterative conception, sets are created stage-by-stage, using as their elements only those which have been created at earlier stages.[8]
>
> In the metaphor of the iterative conception, the steps that build up sets are "operations" of "gathering together" sets to form "new" sets.[9]
>
> ... a set is formed by selecting certain objects ... we want to consider a set as an object and thus allow it to be a member of another set ... When we are forming a set z by choosing its members, we do not yet have the object z, and hence cannot use it as a member of z.[10]

So sets are to be thought of as somehow being formed in stages. This image of sets being built up from below does seem to get at something about what is wrong with paradoxical constructions like Russell's set, but it should also be apparent from the preceding quotations that the nature of the set-forming operation is extremely unclear. This can be seen in the tendency of iterative theorists to put words like "form" in quotes and refer to them as metaphors. Indeed, to speak of "forming" timeless abstract objects in "stages" is, on its face, patently nonsensical. But one searches the iterative literature in vain for a cogent account of what is really meant by this language. The complaint is that "any conviction that the iterative conception may carry is made to depend on metaphorical details that are dismissed as inessential to it".[11]

The difficulty is crucial because the whole point is to understand why things like Russell's set are illegitimate. If the reason is because they cannot be metaphorically "formed" from elements which metaphorically exist "before" they do, and we have no idea what this actually means, then the explanation has little value — negative value, even, because it conveys an illusion of understanding where there is none. It is a little like saying that a ghost is a person who commits malicious acts, only without the person: sets are built up in stages, only without the building and the stages. That is not to deny the intuitive appeal, which I already acknowledged, of the idea that the elements of a set are in some sense prior to the set itself as an explanation of the paradoxes. But what is that sense?

The iterative conception also faces a problem from the opposite direction.[12] Suppose we were to satisfy ourselves that there is some meaningful way of interpreting the idea of "forming in stages" in the atemporal abstract realm of sets. So we could say things like "there is no set of all ordinals because when we undertake the process of forming sets in stages we never reach a point where all ordinals are available to be collected together into a set" and have them mean something. Then how could the "metaphor" of formation be compatible with the basic step of the construction, where we pass from a given stage to its power set, the set of all of its subsets?

At finite stages, there may be no special difficulty. We can simply list all the subsets of a finite set and presume that this suffices to "form" the power set. But we cannot do anything like this for infinite sets. So how would we go about "collecting" all the subsets of, say, the set of natural numbers? Should we in some metaphorical way search through the entire universe of sets, locate all the subsets of N that appear there, and collect them together? No, this answer is completely contrary to the iterative conception; searching through the entire universe of sets is certainly not something we can be allowed to do when constructing a particular stage of the universe. But how else could we perform the task of constructing the power set of N?

Once the set of natural numbers is sitting in front of us, as it were, there is a sense in which its power set is present too. Every subset of N is also there, contained in N. But the power set of N is not available as an articulated object, and getting to that point runs into a circularity issue. Recalling an example from Section 4.3, let R be a relation between numbers and sets, and let $A[n]$ be the formula which asserts of the number n that there exists a set of numbers X to which it bears the relation R. Once the power set of $\mathbb{N}$ is available it may be clear that $A[n]$ has a definite truth value for every value of n, and hence that $\{n : A[\hat{n}]$ is true$\}$ is a well-defined set of numbers. But without having access to the power set of $\mathbb{N}$ we have no way to evaluate this set. It is a set of numbers that becomes available only after all sets of numbers are available. This shows that power sets of infinite sets have a circular quality which could prevent them from being, in any sense, built up from below.

The point is that according to the iterative conception, the set-theoretic universe is thought of as somehow being constructed in an iterative process, yet the basic step in this process is completely nonconstructive: the power set of the preceding level is not "built up" in any sense whatever, it simply appears. And the essentially circular nature of infinite power sets forces this.

6.2 Indefinite extensibility

Self-reproductive processes. Indefinite extensibility. Definite and vague concepts. Circumscription. What does it mean to have a definite conception of a totality?

Let us take a different approach and ask why some concepts have set extensions and others do not. We can restrict our attention to concepts which are known to satisfy the law of excluded middle, i.e., for which we can assert the sentence $\ulcorner(\forall x)(C[[x]] \vee \neg C[[x]])\urcorner$, and which we might therefore presume to have well-defined extensions. The distinction is then between concepts like "being a natural number" whose extension is a set and "being an ordinal" whose extension, if there is one, is a proper class. What is the source of this difference?

A periodically recurring theme throughout this book has been the varying location of the cutoff between sets and proper classes, according to different philosophical stances. For finitists, any infinite collection is effectively a proper class; for predicativists, any uncountable collection is seen this way. What we are interested in here is not correctly situating the cutoff but explaining its nature, wherever it turns out to lie.

Russell wrote of "self-reproductive processes and classes" about which "we can never collect *all* of the terms having the said property into a whole; because, whenever we hope we have them all, the collection which we have immediately proceeds to generate a new term also having the said property".[13] The concept of being a set is a typical example of this phenomenon because, whenever we hope we have them all, what we have is itself a set, X, and then $X \cup \{X\}$ is a larger set all of whose members are sets.

For a finitist, the concept of being a natural number should be seen as self-reproductive, because the "whenever" in "whenever we hope we have them all" only covers situations where we have encountered finitely many numbers: finitists would reject the idea of a possible state of affairs in which an infinite collection had already been canvassed. And given any finite collection of numbers we can "generate" a new number by adding 1 to the largest number in the collection. For predicativists, there is a diagonalizing construction which generates a new real number from any countable set of real numbers. Thus, the general idea of self-reproducing concepts could potentially apply equally well regardless of one's position on the legitimacy of infinite sets.

Russell's expression of his idea contains dubious language about "collecting into a whole" similar to what we saw in the iterative conception. However, Dummett has a reformulation of it which avoids this language. He calls this version *indefinite extensibility*. I will quote Dummett twice because the explanations vary slightly. First,

> a concept is indefinitely extensible if, for any definite characterisation of it, there is a natural extension of this characterisation, which yields a more inclusive concept; this extension will be made according to some general principle for generating such extensions, and, typically, the extended characterisation will be formulated by reference to the previous, unextended, characterisation.[14]

Alternatively,

> an indefinitely extensible concept is one such that, if we can form a definite conception of a totality all of whose members fall under that concept, we can, by reference to that totality, characterize a larger totality all of whose members fall under it.[15]

These passages appeal to our intuitive sense that proper classes are in some way absolutely inexhaustible, and they are more cogent than the iterative conception because what is being "formed" is a conception or characterization of a set rather than the set itself. But their exact meaning is still elusive. In the first one an indefinitely extensible concept is contrasted with a "definite characterisation of it", which seems to suggest that the original concept was ambiguous in some way. It sounds like the process Dummett has in mind here is something like the following. Guided by a vague, informal idea, we produce a sequence of precise formal concepts that incompletely embody the original notion. None of them is definitive, and indeed we have a general principle for converting any given partial formalization into a broader, more inclusive formulation. So there can be no ultimate precise, formal version of the original vague, informal notion. The most we can achieve is an open-ended sequence of partial formalizations.

This interpretation suggests a simple solution to the problem posed above about why only some concepts have set extensions. Perhaps all we have to say is that concepts like "being an ordinal" fail to have well-defined extensions because they are vague?

However, this suggestion does not work. Call a concept *definite* if every individual definitely either does or does not fall under it. *Vague* is conventionally taken to mean the opposite of this, i.e., as implying that there could exist individuals whose status with respect to the concept is undetermined. But that cannot be what it means here because at least some concepts that certainly qualify as indefinitely extensible are clearly not vague in this sense. For instance, according to a finitist the concept of being a prime number is indefinitely extensible: given any finite set of prime numbers we have a finitary procedure (multiply them together, add one, and factorize) which generates at least one new prime number not in the set. Yet there is nothing ambiguous to a finitist about what constitutes a prime number; indeed, we have a finitary, mechanical procedure for testing primality. For a finitist, the assertion that any given number is prime not only has a truth value, it has a decidable truth value. So we have a concept which is both definite in the above sense and indefinitely extensible. Vagueness is not the issue here.

An analogous example can be given in the case of predicativism. Here the concept of being an irrational number would be considered indefinitely extensible because, according to a predicativist, any set of irrational numbers is countable and hence can be diagonalized. But we can check with a computation of countable length whether a given infinite string of digits is eventually periodic, so it is predicatively decidable whether such a string represents an irrational number. Thus, we again have an indefinitely extensible concept whose sense is understood with complete precision. This example is slightly sharper than the previous one because, while the finitist requires successively longer computations to test for primality as the size of the candidate number grows, the predicativist's procedure for testing irrationality always involves a computation of the same countable length. So there is even less room to argue that his grasp of the concept is changing in any way as his repertoire of irrationals grows.

If there is any vagueness in cases like these, it resides not in the application of the concept to any particular individual, but rather in the question of where individuals falling under the concept are to be sought. The finitist is, so to speak, not initially acquainted with all the natural numbers and has no conception of a circumscribed arena in which they appear. The predicativist is situated similarly with respect to the real numbers, with the added feature that he not only lacks initial familiarity with them, he does not even have a clear generating procedure that would potentially produce all of them. Again, the individuals falling under the concept are not to be found in any circumscribed arena.

Now, contrary to what I suggested above, if a concept is "vague" only in the oblique sense of being uncircumscribed then it is not clear that we cannot regard it as having a well-defined extension. But whether we want to say such concepts have extensions does not matter so much at this point because in any case they do not have circumscribed extensions, and this ought to go a long way toward explaining the set/class distinction. The problem is that understanding what "circumscribed" means in this context does not seem all that different from our original goal of understanding how sets differ from classes. We already perceived that sets are limited in some way that proper classes are not, and the image of circumscription does not really add any precision to this thought.

Dummett's second passage quoted above frames the extensibility condition in terms of our being able to enlarge any totality all of whose members fall under the concept. This formulation squares better with the examples just mentioned because there is less of an implication that the underlying concept is evolving, but it also brings out more directly our need to understand what it means to have a "definite conception of a totality". In particular, we must ask how this differs from merely grasping a well-defined concept. (Just how is having a definite conception of the totality of prime numbers any different from knowing what prime numbers are, or to put it differently, once we know how to test for primality, what more do we need in order to be able to say that we have a definite conception of the primes as a "totality"? What is it that the finitist lacks in this case?) This again seems related to, perhaps even essentially identical to, our original question about how sets and classes differ. The definition of indefinite extensibility evidently presumes that this question has already been answered. We now see that indefinite extensibility cannot be used as a criterion for differentiating sets and classes; to the contrary, we already need to be able to differentiate these concepts before we can make sense of indefinite extensibility.

6.3 Surveyable concepts

Infinite computations versus infinite sets. Surveyability. Decidability. Truth values. Decidability does not entail surveyability. Resolution of the set-theoretic paradoxes. Surveying power sets. Correcting the iterative conception and indefinite extensibility. Surveying power sets.

We are still unclear about the distinction between sets and proper classes. I find it helpful to relate this question to the issue of infinitely long computations.

Although there have been places in this book where we had to assume that infinitely long computations are theoretically possible, I tried to indicate where this assumption was needed and to avoid adopting an official position on the matter. In fact, I do think there is a meaningful sense in which they are possible, and I will come back to this point at the end of this section, but I do not insist on it here. For the sake of the present discussion we can be pluralistic on the issue.

Finitists do not agree that infinitely long computation — computations which do not merely continue indefinitely, but which return a result after having executed an infinite series of steps — are possible, even in principle. Finitists also do not agree that infinite sets exist, and the two positions are obviously connected.

Predicativists accept computations with a countably infinite number of steps, but not uncountably long computations, and they likewise reject uncountable sets. Again, the two positions are connected. In platonism the analogous comparison would be between proper classes and computations of proper class length. The latter came up in Section 4.2, and I tried to show there that they also do not make sense from a platonist perspective.

We are led to the idea that a characteristic feature of a set that distinguishes it from a general collection is the theoretical possibility of performing a computation over it. Thus, let us say that a concept is *surveyable* if it is possible, in principle, to exhaustively survey all of the individuals which fall under it. That is, in principle, we could perform to completion the task of examining, one at a time, all the individuals falling under the concept.

Surveyability is closely related to the notion of decidability. A concept is said to be *decidable* if there is a mechanical procedure which can determine whether any given object falls under it. For instance, primality is decidable because, given any natural number, we can systematically divide it by each smaller number, and therefore determine after a finite number of steps whether the original number has any exact divisors besides 1 and itself. In this instance the answer will be found after a finite computation, and decidability is usually understood against an implicit background of finitism. However, since we have also been concerned with infinitely long computations, I have taken care to use the term "finitely decidable" for concepts which can be mechanically tested in a finite number of steps.

In contrast, in Section 5.2 I mentioned the possibility of mechanically evaluating the truth values of arbitrary arithmetical sentences by means of countably long computations. So these truth values may be considered to be predicatively decidable, but not finitistically decidable. It is worth noting here that the connection to the problem of determining truth values is a general feature of surveyability: the truth values of sentences which quantify over some collection are decidable, in the generalized sense, if computations, in the generalized sense, can be performed over the collection, i.e., if the collection is surveyable. (That is, assuming that we are talking about sentences all of whose atomic subformulas are decidable.) If the range of the variables is surveyable then in principle we should be able to evaluate the truth of quantified statements by direct inspection.

Since terms like "decidable" and "computable" are usually meant in a finitistic sense, I must emphasize that here we are using them in a general sense which could allow infinite or even uncountable processes. Clearly, the question of what is possible in principle is open to interpretation, with different foundational stances taking widely varying positions, and even being largely characterized by these positions. So the idea of surveyability cannot be fully cashed out until we settle on one of these views over the others. But even before we do so, it already seems more informative than ideas of "circumscription" and "totality". According to a platonist, the collection of natural numbers is a circumscribed totality, while according to a finitist it is not. What does this mean? It means that the platonist thinks that a computation whose steps are indexed by the natural numbers in principle could be run to completion, while a finitist does not.

The example of primality shows that merely being able to determine whether any given object falls under a certain concept does not entail that the concept is surveyable. At least, it shows that finitistic decidability does not imply finitistic surveyability. It is not hard to give examples that show the same thing with regard to predicativism or platonism. Indeed, the example of irrationality mentioned in the last section illustrates this point in a predicative setting. Platonically, the concept of being, say, an ordinal number with countable cofinality is decidable in the generalized sense of being testable by means of a computation which, given any ordinal as input, terminates after an ordinal number of steps. But it is not surveyable; the collection of all ordinals with countable cofinality constitutes a proper class. So it is a general fact that decidability does not entail surveyability.

Looking at the matter in computational terms helps to clarify distinctions like these. For instance, someone not previously familiar with the set-theoretic paradoxes might find it difficult to gauge the meaning of the question

> If we can decide whether any given individual falls under some concept, does it follow that all the individuals falling under that concept constitute a definite totality?

and would probably be inclined to answer "yes". Phrased in terms of surveyability, the analogous question is

> If we can decide whether any given individual falls under some concept, does it follow that we can exhaustively survey all the individuals falling under that concept?

and this is not only more lucid, it also does a better job of locating the burden. The default answer is clearly that being able to diagnose whether any given individual has some property need not entail that we are able to inventory all the individuals with that property; ergo, there are, on the face of it, (at least) two distinct kinds of concepts: those which are surveyable and those which are merely decidable. This moves us toward an explanation of the difference between sets and classes.

In particular, a resolution of the set theoretic paradoxes is available to us now as a consequence of the premise that sets are surveyable collections, not merely collections. The collections appearing in the standard paradoxes (the collection of all ordinals, the collection of all sets, etc.) are not truly paradoxical because they are not actually sets, and they are not sets because they are not surveyable. If there is any question as to what could render us unable, even in principle, to survey all the sets there are, we just have to observe that being in the position of having completed a survey of all the sets there are is not a possible state of affairs. This is where the notion of self-reproduction or indefinite extensibility can be enlightening. No matter how many sets we have managed to survey, the collection of all the sets we have surveyed will necessarily be a set that we have not yet surveyed. Thus, even in principle there is no way we could survey all the sets there are.

So there are concepts which we can be sure are not surveyable. We may attribute the set theoretic paradoxes to a mistaken implicit assumption that all concepts are surveyable; once this assumption is denied these paradoxes evaporate. (Of course, this still leaves us with the analogous paradoxes at the level of classes. I will discuss that topic in the next section.)

It could be objected that I have merely traded one mystery for another. Certainly, the idea of performing a computation over a set carries its own problems, although for finitists and predicativists, at least, these problems do not seem so great. But I want to insist that even for platonists the idea of surveyability has clarifying value. The basic problem for platonists is to reconcile the idea that a set is nothing more than "a collection of objects"[16] and the principle that sets are themselves objects with the fact that the universe of sets fails to be a set. The iterative conception attempts to resolve this by characterizing sets in terms of some sort of metaphorical building process that is described using temporal language, which absurdly makes it sound "as if sets were continually being created".[17] Framing the discussion not in terms of forming sets, but of computing over them, avoids this problem because computation can legitimately be thought of as a temporal process.

Similarly, the basic challenge to indefinite extensibility is to explain what is being indefinitely extended. It cannot be the concept itself, because we can give examples for various foundational stances (finitism, predicativism, platonism) of perfectly definite concepts (primality, irrationality, countable cofinality) which are also indefinitely extensible. But if what we are extending is "totalities" which are subordinate to a concept and which can be sets but cannot be proper classes, then we need to already understand what a "totality" is, i.e., we need to already understand the difference between sets and proper classes. And this amounts to already being able to distinguish ordinary concepts from indefinitely extensible concepts. The clarification that surveyability provides here is to think in terms not of extending totalities but of completing a survey.

The characterization of sets in terms of surveyability has consequences for some of the topics discussed earlier in this book. In Section 2.3, for instance, we constructed a truth function for any language that is interpreted in a set model, but we qualified it with the warning that one had to accept the idea of computations over arbitrary sets. According to the conception of sets considered in the present section, this qualification is unnecessary: by definition, truth functions for languages with set models are in principle computable.

In Section 4.5 I introduced a "($\forall$) law" which was supposed to be valid "in settings where the variable x has sufficiently limited range that proofs of all instances of $A[\hat{x}]$ could be collected together to get a single proof of $(\forall x)A[x]$". We can now simplify this condition to say that the ($\forall$) law is valid when the relevant variable ranges over a set.

In particular, I included the ($\forall$) law in the formulation of PA^A, and it was noted in Section 5.2 that this inclusion should be justified if we are willing to accept countably long constructions. If the characterization of sets in terms of surveyability is granted, then a simpler way to say this is that the system PA^A should be regarded as valid if one considers $\mathbb{N}$ to be a set. That is, it is predicatively valid, but not finitistically valid.

Similarly, the inference mentioned in Section 4.6 to the assertibility of $(\forall A)T(\ulcorner A \to A \urcorner)$ for any classical truth predicate T defined for some object language requires only that the object language contain a set, not a proper class, of sentences.

I can also now clarify the power sets problem for the iterative conception which I discussed at the end of Section 6.1. Instead of talking about the difficulty of grasping the power set of $\mathbb{N}$ as an "articulated object", we should ask about the possibility of running computations over it. In other words, the question is how we go from being able to run computations over a set X to being able to run computations over its power set.

My view is that we have a naive intuition for the surveyability of the natural numbers — specifically, a sense of their sequential availability — that we do not have for the power set of $\mathbb{N}$. I can imagine a computation whose speed increases exponentially as it progresses, such that the result of an infinite sequence of computational steps is returned after a finite interval of time.[18] But I have no sense of what it would be like to run to completion a computation which involved a separate step for every subset of $\mathbb{N}$. So that makes me a predicativist, but I only mention this in passing as I do not think it bears directly on the central issues of this book.

## 6.4	Abstract objects

Nominalism and realism revisited. Concrete proxies. Modal nominalism.
Explicit lists as proxies for sets. Definite predicates as proxies for classes.

One could ask whether our main criticism of the iterative conception,
that it makes no sense to speak of "forming" abstract objects such as sets,
also applies to the idea of surveyability introduced in the last section. Is
it cogent to talk about "surveying" an abstract object? How would one do
this?

The short answer is that we survey the natural numbers by surveying
some concrete representations of them, such as their decimal representations. But in order to develop this answer I need to say something about
the general problem of abstract objects.

As I mentioned in Section 1.2, the basic opposition here is between
nominalism and realism. According to nominalism, there are no abstract
objects, and talk about abstract objects is meaningful just to the extent
that it is really talk about concrete objects in disguise. Whereas realism
holds that abstract objects genuinely exist in some substantial sense.

In its strongest form, realism says that every meaningful noun phrase
(excluding idioms and fictional discourse, maybe) refers to an object. So
if we can meaningfully say that 37 is prime, then there must be such a
thing as 37. And since we cannot locate 37 as a specific concrete object in
our physical world, it must therefore be an abstract object which is located
elsewhere. One can form a crude picture of a universe of abstracta which
enjoys some kind of existence parallel to our physical world and in which the
abstract objects somehow reside. But this picture is straightforwardly self-
defeating because, if we can talk about it, then this abstract universe would
itself have to be an abstract object, leading to the usual kinds of paradoxes.
So one has to replace the crude picture with something more sophisticated.
But as one gets away from the idea of a literal abstract universe literally
populated with abstract objects, it becomes less clear what actual content
there is to the assertion that abstract objects exist.

There is a large philosophical literature on the problem of abstract objects. How we, as physical beings, could know anything about non-physical
objects, has been identified as a key difficulty.[19] But at some level one feels
there is no serious issue here. We know perfectly well how to calculate with
numbers, how to check whether 37 is prime, for example. Whether 37 is a
real thing, what this assertion means, and how we would know that it is

true, are no doubt interesting questions, but it is not so clear that anything mathematically important hinges on their answers.

A large part of the explanation of how we get information about abstract objects must be through our interactions not with these alleged objects themselves, but with various kinds of concrete proxies for them, such as the numeral "37" which actually appears in calculations. Our immediate knowledge is about these proxies, not the abstract things they are supposed to represent. But then, if the concrete proxies do all the work, then perhaps we have no particular need to include the abstract objects they allegedly represent in our vision of the world. This could be the beginning of defense of nominalism.

(Are numerals themselves concrete? Well, they are clearly more concrete than the abstract numbers for which they can go proxy. But syntactic objects can themselves be represented by physical proxies like configurations of ink or chalk which are even more concrete. There could be a hierarchy of abstractness, with no sharp line dividing abstract from concrete.)

Of course, the idea of reducing everything to the level of proxies only works if the proxies themselves exist. If there were no abstract numbers, only concrete physical representations of numbers, and the physical world happened to be finite, then we would be wrong to think of the set $\mathbb{N}$ as infinite. But my sense here is that what happens to be the case in our physical world is irrelevant to the global truths of mathematics and logic. These truths should hold good in any possible world. So if "there are infinitely many primes" is to be understood nominalistically as a fact about certain numerical calculations, say, then these should include calculations that might take place not just in our world, but in any possible world. One then gets into a version of nominalism according to which mathematical truths are not truths about some particular abstract universe of mathematical objects, but truths about arbitrary possible worlds of concrete objects.[20]

This kind of possible-worlds nominalism seems like the right setting for the idea of surveyability. We are interested in what kinds of computations can be performed in principle, and perhaps another way to get at that is to ask what computations can be performed in some possible world. On the surveyability account, the set $\{1, 2, \ldots, 100\}$ is a set and not a proper class because we can, in fact, run computations over the numbers 1, 2, ..., 100, and whether $\mathbb{N}$ is a set or a proper class turns on whether there is a possible world in which computations can be run over the entire infinite sequence 1, 2, ... This suggests that the idea of surveyability can be made sense of in some kind of nominalistic terms.

But then, maybe the idea of forming a set, which I discussed in Section 6.1, can be given a similar interpretation in terms of proxies. If a numeral can go proxy for a number, what analogous concrete thing can go proxy for a set? One obvious answer is, an explicit list of names of objects. On the face of it, an explicit list of names is a perfectly natural concrete proxy for a set. (A list of names, not a list or arrangement of the objects themselves. We have to distinguish between the object x and the set $\{x\}$, for instance.) This might let us make sense of words like "form" and "collect" in the context of sets; we can certainly form lists of names. So the idea could be to think in terms of forming proxies for sets, not the sets themselves (which, once one is deliberately thinking in terms of concrete proxies, tend to fade in significance).

The "explicit list of names" approach does suffer from the complication that different lists can identify the same set — the names could be rearranged, for example, or there might be objects which have more than one name. But this complication does not seem that great. We simply have to consider two lists to be "the same" if they name the same objects, irrespective of order. That is, we need to adopt the right identity criterion. Nontrivial identity criteria are unexceptionable; they are used even in the case of physical objects. I consider my desk now to be the "same" thing as my desk five minutes ago, despite the fact that the two objects occupy different locations in spacetime.

A more serious complication is that working with names of objects seems to imply a fixed background language, when to the contrary, one supposes that a richer language which is able to name more objects can always be produced. So in order to handle the nested hierarchical nature of the universe of sets we presumably need the language we are using to have an open-ended ability to add new names.

In any case, we should no longer feel threatened by the set-theoretic paradoxes. If we take a nominalistic approach and think of sets in terms of some kind of concrete proxies, then the nonexistence of a set of all sets is explained by the fact that any system of proxies can be extended further. But on a realist account there is also no problem: there may be a collection of all sets, but it is not a set because it is not surveyable.

What about classes? Can we say that they exist? It is clear that if a concept is definite but not surveyable then we cannot, in Dummett's words, "form a definite conception of a totality" consisting of precisely those individuals which fall under it. But that is just because its extension is not a "totality", i.e., is not surveyable, not because its extension is in any sense indefinite. Confusion can arise here from equating "totality" with "definite collection", since this implies that collections which are not totalities cannot be definite, and hence that it must somehow be possible for definite concepts to have hazy extensions. But once the distinction afforded by the notion of surveyability is available there is no need to make this equation.

To the contrary, it is not clear what is wrong, if we are finitists, with regarding the natural numbers as comprising a well-defined albeit unsurveyable collection, or if we are predicativists or platonists, with regarding the real numbers or the ordinals in the same way. Indeed, it is not obvious what more could be reasonably demanded of a well-defined collection than that every object should definitely either belong or not belong to it.

However, if classes are understood realistically then it is hard to see why the concept of being a class should not itself be definite. This cannot be right, because it would imply that there is a class of all classes. The nominalist account in terms of proxies does better here. Since we cannot explicitly list all the elements of a proper class — by definition, such a list could never be completed — we have to use something less direct as a proxy for a proper class. The natural choice here is to represent a class by a predicate, i.e., a formula with one free variable, which characterizes those objects which belong to the class. Not every formula with one free variable is suitable, however. We have seen that we have to abandon the law of excluded middle when discussing arbitrary concepts. So if $A[x]$ is any formula with one free variable, we may not be able to assert $(\forall x)(A[x] \vee \neg A[x])$, and this means that we cannot assume that every formula with one free variable represents a class to which every object definitely either does or does not belong. If classes are to have this character of definiteness, then we have to use as proxies only those formulas that have the same definite character. And here is our solution to the class-theoretic paradoxes: there is no class of all classes because the concept "formula for which $(\forall x)(A[x] \vee \neg A[x])$ is assertible" is, apparently, not definite.

The situation is very different with sets. Since sets are surveyable, we are able to use proxies for them which are simply explicit lists of their elements. These lists themselves are surveyable, and we should therefore be able to decide by direct inspection whether a given object is or is not a proxy for a set. This supports the idea that the concept of being a set is definite, and that there is therefore a class of all sets. In contrast, proxies for classes have to characterize the classes they represent in some indirect fashion, not by explicitly listing their elements, and this creates the possibility of there being objects whose status as a proxy for a class cannot be determined. So when classes are approached via proxies, the concept of being a class is apparently not definite, which means there is no class of all classes.

Thus, we have three levels of paradoxes, and they are defeated in three different ways. There is no set of all sets because the collection of all sets is not surveyable. There is no class of all classes because the concept of being a class is not definite. And while there is a concept of being a concept, in this setting there are good reasons not to adopt the law of excluded middle and the release law, which suffices to defuse the paradoxes.

This analysis seems to me to show that a realist account of proper classes as legitimate objects is not possible. The only way it makes sense to say that "being a class" is not a definite concept is by interpreting talk about classes in terms of concrete proxies which might or might not play the role of fictional abstract classes, and for which there could be cases where this is not determined. If one insists on dealing with the alleged classes directly then it is hard to see how they could be indefinite. This does not preclude a mixed account, according to which, say, sets are genuine abstract objects while proper classes are understood nominalistically.

It also seems worth emphasizing that the resolution of the class version of Russell's paradox depends essentially on the role that intuitionistic logic plays in the context of reasoning about general concepts. It is the absence of the law of excluded middle that opens the possibility of there being indefinite concepts, and the consequent possibility that the concept of being a class might not have a definite extension.

6.5   Conclusions

I summarize the main points of the book.

There are good reasons to think that we have a naive concept of truth. First of all, we have a conviction that when we say some sentence is true, or some sentence is the case, or one sentence implies another (see p. 18), we are making unrestrictedly meaningful assertions. We also have an intuition for concepts like "proposition", "concept", and "relation", and relations like "the object x falls under the concept C", all of which require the existence of a concept of truth (see p. 42).

But Tarski got truth profoundly wrong. His characterization of truth only makes sense as applied to a target language within some metasystem, which is not consistent with our intuition. This is most powerfully seen in the total inability of a Tarskian notion of falling under to let us make global statements about concepts. There is no way to say "for any concept C, if 0 falls under C then 0 falls under C'" in a Tarskian sense of falling under and have it apply to every concept expressible in the language in which it is said (see Section 2.7).

Besides that, there is the problem of saying what it means to be a Tarskian truth predicate. Our intuition tells us that an interpreted language is the kind of thing a truth predicate should apply to, but we cannot say what a truth predicate for an interpreted language is without invoking truth (the Tarskian catastrophe; see Section 1.3). Tarski's evasion of this difficulty by having truth predicates operate between formal systems and metasystems, while ingenious, destroys the functionality of truth. It is indifferent to the core feature of truth, which is its global disquotational property. Having the T-scheme only as a scheme brings no deductive power (see Sections 2.5 and 3.7).

Tarski's construction of a truth predicate for an arbitrary language interpreted in a set model (see Section 2.3) is more successful — it has all the properties we want and it applies to the right kind of thing, an interpreted language. But it only works for set models. It does not apply to a language in which the word "set" is interpreted to range over all sets, for instance. If variables can range over all sets then the language is not being interpreted in a set model and the truth predicate construction does not apply to it. One could try to extend the construction to languages with class models, but this leads to the same difficulty: if we interpret the word "class" to range over all classes, then our language cannot even have a class model. It is not an option to dismiss such languages as frivolous or inessential. In particular, the truth predicate construction can never apply to the very language in which it is described. To reject languages to which the construction does not apply is to reject any language in which we are able to express the fact that the construction has the properties we want of a truth predicate. This was the first example we saw of the general "revenge" phenomenon in which the functioning of language is explained using concepts which cannot be spoken of in any language to which the explanation applies (see Section 2.6).

I have not spent much time on Kripke's celebrated theory of truth, because it is really a theory not of truth but of grounded truth. For that matter, it is not even a theory, it is a construction — there is no general characterization of what constitutes a grounded truth predicate. And like

Tarski's construction, it cannot be applied to the language in which it is presented.

The problems discussed above are intractable because the basic idea of characterizing truth using the T-scheme is bad. The misfortune is that this bad idea is, on its face, so attractive. This leads us to expect it to perform a function that it is unable to perform. It would be better to introduce a neutral term and speak of, say, "disquotational predicates", not truth predicates, in relation to the T-scheme.

It makes more sense to regard assertibility as realizing our naive notion of truth. Assertibility is the genuinely global concept — to say that a sentence is assertible really means something (see Section 4.1). We can use assertibility to make sense of all the things (implication, propositions, falling under, etc.) that we could not make sense of on a Tarskian approach (see Section 4.6). Even the Tarskian catastrophe can be averted: we can finally say what it means for the T-scheme to hold for a class of sentences.

The intuitive appeal of the T-scheme might be explained in terms of an equivocation between restricted and unrestricted versions of assertibility. The intuition for the implication $\mathbb{T}(A) \rightarrow A$ may come from thinking of $\mathbb{T}(A)$ as expressing that A can be asserted on the basis of some limited form of reasoning which is known to be sound. In other words, there could be an equivocation in informal settings between $\mathbb{A}$ and some $\mathbb{D}_S$ (see p. 109).

In Section 2.4 I argued that our pretheoretic conception of truth is not consistent. But maybe our pretheoretic conception is, after all, a legitimate conception of assertibility, and we misunderstand it, thinking that we have a right to affirm the release law when we really do not. This would reverse a central conclusion of the first part of the book. We *do* have a valid pretheoretic concept of truth, but it is not characterized by the T-scheme. The relase law only works as a deduction rule, not as an implication, and the law of excluded middle cannot be assumed.

The nature of assertibility is peculiar. It must be reasoned about constructively, which requires a conceptual shift. An analogy with physics might be helpful here. According to standard quantum mechanics, a particle's position and momentum cannot be simultaneously measured with arbitrary accuracy. Since it is hard to imagine a particle that does not actually have exact position and momentum coordinates, a natural reaction is to suppose that these values exist even if they cannot be observed — the "hidden variables" interpretation. One might think that this assumption would be harmless. Could there be any testable consequences of assuming the existence of *unmeasurable* position and momentum values? This assumption would remove much of the mystery of quantum mechanics; why not make it?

Surprisingly, Bell showed that even if hidden variables cannot be measured, their mere existence implies (under very mild assumptions) that in some situations there will be statistical correlations between distant measurements which contradict the predictions of quantum mechanics.[21] The lesson is that we are not always free to assume the existence of quantities that cannot be measured.

There is no direct connection between these physics issues and the topics discussed in this book, but I see a similarity in relation to the question of whether we may assume that things we cannot measure nonetheless exist. In quantum mechanics, at least sometimes we apparently are not free to make that kind of assumption. I claim that the semantic paradoxes show that in the realm of pure logic, we similarly are not free to generally assume the existence of truth values which in principle cannot be evaluated.

This issue arises when we quantify over proper classes. If a proper class is by definition a collection over which computations cannot be performed to completion (see Section 6.3), then we may be in principle unable to evaluate the truth of a sentence of the form "every element of the class X bears the relation R to another element of X", for example. It is natural to assume that such sentences nonetheless have well-defined truth values, even if possibly no one could ever know what they are. But there is a circularity concern here relating to the indefinitely extensible nature of proper classes. Since we can never be in a position where we have completed a survey of X, might answers to questions of the preceding type, if known, be used to build more of X in a way that would cause the answers to change?

This is the sort of issue we face with assertibility. If we assume that the variant assertible liar sentence (see p. 123) has a definite truth value — that is, we assume $L_2 \vee \neg L_2$ — then we can show that a falsehood is assertible. This is not quite as bad as proving a falsehood outright, but it is still a disaster, as it would entail that everything is assertible. Since both L_2 and $\neg L_2$ lead to this conclusion, we do not have to know which of them obtains, only that one of them does. The mere existence of a presumably unknowable truth value is enough to cause problems.

One has to be careful here because there is now a temptation to conclude that the law of excluded middle fails for L_2. That inference would be valid if we knew $\mathbb{A}(\bot) \to \bot$, i.e., no falsehood is assertible. But this assumption too has circularity issues. If we affirm that our reasoning is consistent and then use that affirmation to make further deductions which fall under the scope of the original affirmation, then that affirmation acquires a self-referential quality which renders it illegitimate. We may hope and expect that no falsehood is assertible, but we cannot mentally jump to a nonexistent future point at which all questions have been answered and no falsehood has been proven, and then use that information to prove theorems now. We have to leave some questions open and we have to accept that unknowable truth values may not exist.

This view touches on dialetheism, the idea that there are true contradictions. There are few dialetheists: most people believe that contradictions cannot be true. I do not believe that contradictions can be true, a slightly different position. My position leaves open the possibility that they could turn out to be true after all. More precisely, I do not believe that $0 = 1$ is assertible, but I do not assume $\neg A(\text{"}0 = 1\text{"})$. However, that is the farthest we can get from dialetheism. The positive affirmation that $0 = 1$ is not assertible is self-defeating, because, via the assertible liar paradox, it entails that $0 = 1$. Given that $0 = 1$ cannot be asserted, we can infer that the sentence that asserts of itself that it cannot be asserted, indeed cannot be asserted. And thus we have proven that sentence, which means that it can be asserted, a contradiction.

We may be said to skirt dialetheism, in that $\neg\neg A(\text{"}0 = 1\text{"})$ is a provable theorem. As I just explained, this is the best we can do. Perhaps saying that "$0 = 1$" is not not assertible might not seem so far from saying that "$0 = 1$" is assertible. However, that is sloppy thinking. The correct interpretation is that assuming "$0 = 1$" is not assertible leads to a contradiction, which it does, and given that we are reasoning constructively, this is only weakly related to the statement that "$0 = 1$" is assertible. If "$0 = 1$" is assertible then it is not not assertible, but we have no right to draw the reverse implication.

We have to walk a logical tightrope. The only way to keep our reasoning consistent is to leave open the possibility that it might turn out to be inconsistent.

But this does keep our reasoning consistent, as I showed in a variety of settings in Chapter 5. We evade not only the assertibility version of the liar paradox, but also the assertibility version of Russell's paradox. As an unexpected bonus, we find that in arithmetical settings assertibility reasoning allows us to, in a way, bypass the second incompleteness theorem and formulate extensions of Peano arithmetic which effectively affirm their own soundness and consistency. (See Section 5.2.)

The improvement in second order logic is even greater, because without assertibility we have no ability whatsoever to reason globally about concepts. The source of the difficulty is that unrestricted quantification over schematic variables makes no sense. So in order to quantify over concepts, to make general statements about concepts, we need to use ordinary variables which range over concepts, but these have no assertoric force. We cannot then combine a concept variable with an object variable to make a sentence, only to make a term. We need to use an assertibility predicate to give these terms assertoric force. Thus, the way to say that two concepts C and D are equal is not to say "for all x, $C[[x]]$ if and only if $D[[x]]$", because that is not a grammatical statement. We have to say "the sentence $\ulcorner$for all x, $C[[x]]$ if and only if $D[[x]]\urcorner$ is assertible" (with quasi-quotation marks, a crucial difference).

This is the correct form of Frege's notorious Basic Law V. Since concepts are objects, we do not need to introduce their extensions as separate objects. But we do need to introduce a device to make them assertoric, as we did just above. Another approach would be to only consider concepts that have set extensions, and work directly with these extensions. This leads to set-theoretic systems. But the limitation of this approach is that not every concept expressible in the language has a set extension, so that we then need to introduce more complicated rules for deciding which concepts give rise to sets. In order to reason about all concepts, including those which do not have set extensions, assertibility is needed.

All this should explain my reasons for saying in the Preface that assertibility is essential for the resolution of a variety of philosophical problems that are classically intractable.

Notes

Chapter 1

1 (p. 2). [74], p. 7.

2 (p. 3). [52], p. 185 in [7].

3 (p. 5). E.g., "I can no more understand the suggestion that a sentence might be true otherwise than in virtue of its expressing a true proposition than I can understand the suggestion that a name might be honorable otherwise than in virtue of its being borne by an honorable individual or family" ([94], p. 283). A similar sentiment of incredulity is expressed in [80], p. 512 ("What can it be," etc.).

4 (p. 5). [33], p. 292.

5 (p. 9). See [87] or [88] for an account of his theory. [87] is the more technical version.

6 (p. 9). Frege already had this idea: "the sentence 'I smell the scent of violets' has just the same content as the sentence 'it is true that I smell the scent of violets'" ([33], p. 293).

7 (p. 9). The propositional version is "The proposition that $\mathcal{A}$ is true if and only if $\mathcal{A}$".

8 (p. 11). [102], p. 150, note 12.

9 (p. 11). [88], p. 359.

10 (p. 12). [75], p. 33.

11 (p. 12). [76], p. 67.

12 (p. 12). Some would argue that *substitutional quantification* can also achieve this goal. I discuss this in Section 2.5.

13 (p. 13). [76], p. 12.

14 (p. 13). A similar complaint is made in [72], p. 319.

15 (p. 13). Even Tarski appears to commit this error at one point, saying an instance of the T-scheme must "hold" ([88], p. 344).

16 (p. 18). Tarski arguably makes this error as well ([88], p. 350).

17 (p. 27). [26], p. 118.

18 (p. 27). [53], p. 68.

19 (p. 28). [18], p. 7.

20 (p. 29). [11].

Chapter 2

1 (p. 31). "Attributes are to predicates, or open sentences, as propositions are to closed sentences" ([76], p. 67).

2 (p. 31). [32].

3 (p. 31). [31].

4 (p. 32). This point has been made multiple times ([76], p. 28; [83]; [107], p. 255).

5 (p. 32). I omit mention of Frege's paradox of the concept *horse*. This was a difficulty that Frege famously fell into as a result of his assumption that predicates refer to concepts. The matter is discussed fully in [83] and [107].

6 (p. 38). The general idea of truth in a model goes back to [90].

7 (p. 41). [103], introduction.

8 (p. 43). E.g., [13], pp. 275-276: "the freshman fallacy that demands that we *define* our terms", "the same ugly urge to define", etc.

9 (p. 44). [86], p. 46.

10 (p. 45). [53], p. 120.

11 (p. 46). [87], p. 273.

12 (p. 48). [53].

13 (p. 49). Similar points are made in [44], p. 73 and [72], p. 319.

14 (p. 49). [53], p. 137. This follows a similar suggestion by Tarski ([87], p. 259).

15 (p. 51). [29], p. 259.

16 (p. 52). Cf. [76], p. 12 or [21], p. 226.

17 (p. 52). Cf. [76], p. 67; the same point is made by many authors.

18 (p. 52). As argued in [94].

19 (p. 52). [87], p. 160.

20 (p. 52). [14], p. 135, note 17.

21 (p. 52). For instance, [53], p. 25.

22 (p. 53). [104], p. 3.

23 (p. 54). [10], p. 33.

24 (p. 54). [10], p. 46.

25 (p. 54). [10], p. 46, note 22.

26 (p. 55). [32], p. 54.

27 (p. 55). [103], 6.522.

28 (p. 55). [63].

29 (p. 55). [63], p. 221.

30 (p. 55). McGee is aware of this argument and gives his response on p. 222 of [63].

31 (p. 55). [45].

32 (p. 56). [45], p. 256.

33 (p. 56). [1].

34 (p. 56). [1], p. 130.

35 (p. 56). [1], p. 173.

36 (p. 57). [58].

37 (p. 57). [58], p. 714.

38 (p. 57). [86].

39 (p. 58). [86], p. 180.

Chapter 3

1 (p. 62). [35].

2 (p. 66). Essentially due to Gödel [36].

3 (p. 66). [49].

4 (p. 69). See [91].

5 (p. 70). [16], [66].

6 (p. 72). The Church-Turing thesis was first explicitly formulated by Kleene [56].

7 (p. 75). [37].

8 (p. 76). [87], p. 273.

9 (p. 83). [88], p. 346.

10 (p. 85). Tarski addresses this criticism in [87], p. 257. His response is to invoke an infinitary deduction rule, essentially the same idea that was later proposed by Horwich, as we discussed in Section 2.5.

11 (p. 86). I take this to be the sort of difficulty that Putnam is getting at with his comment about ω-inconsistency in [72], p. 41.

12 (p. 86). [46], p. 19.

Chapter 4

1 (p. 90). [52].

2 (p. 92). [19], p. 184.

3 (p. 93). E.g., [105], p. 34: "The proliferation of independence results in set theory certainly undermines any notion of a canonical universe of sets. When the generalization of Cohen's method to models of number theory is developed, then this will also be true of number theory."

4 (p. 95). [22], p. 224/106.

5 (p. 95). [51].

6 (p. 95). For example: "Voici donc l'affirmation brouwerienne de p: On sait démontrer p" ([50], p. 959); "An assertion is on this view simply to be understood as a claim to the effect that a proof of the asserted sentence is known" ([69], 46); "The assertion of $A \vee \neg A$ is therefore a claim to have, or to be able to find, a proof or disproof of A" ([24], p. 21).

7 (p. 95). [57], p. 201.

8 (p. 95). Similar comments are made in Section 4 of [69].

9 (p. 96). Sometimes both errors are committed simultaneously: "an assertion of such a statement is to be construed, not as a claim that it is true, but as a claim that a proof of it exists or can be constructed" ([23], p. 70).

10 (p. 96). [23], p. 46.

11 (p. 97). [62], p. 55.

12 (p. 97). Indeed, occurring in this form in [27], p. 253: "The primary role of proof in mathematical discourse is that it serves to justify our treating the theorem proved *as true* and asserting it *as true*" (italics added). And again on p. 258.

13 (p. 97). In its starkest form: "a proposition is defined by laying down what counts as a proof of the proposition" ([60], p. 11).

14 (p. 98). This difficulty is noted by Weinstein ([101], p. 263), who attributes the observation to Kreisel.

15 (p. 98). Díez proposes that we take a proof of $(\forall x)A[x]$ not to be a construction of a family of proofs, but a single proof that contains x as a free variable. Thus, the meaning of $(\forall x)A[x]$ is directly given by the introduction rule for $\forall$ in natural deduction. But as he immediately notes, this generates a universal quantification issue for the disjunction clause, because once "free-variable" proofs are in play we cannot say that a free variable proof of $A[x] \vee B[x]$ is something which is either a free variable proof of $A[x]$ or a free variable proof of $B[x]$. We have to allow it to be a proof of $A[\hat{x}]$ for some values of x and a proof of $B[\hat{x}]$ for other values of x. So we lose the ability to recognize proofs of disjunctions ([17], p. 420).

16 (p. 100). [55].

17 (p. 102). The idea that quantifying over proper classes may call for intuitionistic logic is one of Dummett's most interesting contributions [22].

18 (p. 102). Dummett suggests a picture "of objects springing into being in response to our probing. We do not *make* the objects but must accept them as we find them ... but

they were not already there for our statements to be true or false of" ([19], p. 185). This characterization may be helpful provided one understands that it is "of course intended only as a picture".

19 (p. 103). [106], p. 5.

20 (p. 105). Cf. [3].

21 (p. 106). This is Dummett's *principle K* [23].

22 (p. 106). A form of *Moore's paradox* ([104], p. 190). Gödel objects that "one may conjecture the truth of a universal proposition ... and at the same time conjecture that no general proof for this fact exists" ([40], p. 313). But it is unclear what significance this comment has since one can (reasonably) conjecture false statements. More pointedly, one can formulate the conjecture that some proposition is true but not provable. However, this conjecture has the strange quality that we are apparently able to establish that it could never be proven. So its significance is again unclear.

23 (p. 106). "The assertions '⊢ p' and '⊢ $+p$' [p is provable] have exactly the same meaning" ([51], p. 60). "To assert is to prove: $\phi \leftrightarrow \exists p(p$ is a proof of $\phi)$" ([4], p. 36, stated with reservations). "A is true if and only if A can be proved to be true" ([61], p. 38).

24 (p. 107). Wright formulates a notion of *superassertibility* in these terms: "A statement is superassertible, then, if and only if it is, or can be, warranted and some warrant for it would survive arbitrarily close scrutiny of its pedegree and arbitrarily extensive increments to or other forms of improvement of our information" ([106], p. 48). This improves on a similar formulation of Putnam [71] in that it does not assume the existence of "epistemically ideal conditions", only, one might say, arbitrarily good approximations to a fictional ideal state.

25 (p. 109). See [70].

26 (p. 110). [43], [57].

27 (p. 110). See, e.g., [91], p. 9, or [61], third lecture, or [4], p. 39.

28 (p. 110). There is some controversy over whether the clause about being able to recognize that the construction succeeds needs to be articulated. It seems superfluous, since according to intuitionism saying that a construction has the desired property already implies that it can be recognized to do so. The point still stands that proving A is importantly different from proving that some construction generates a proof of A. See also the discussion of proofs of $\forall\exists$ sentences near the end of this section.

29 (p. 111). Previous axiomatizations are extremely rare. I only know of one full attempt, in [38], where assertibility is effectively formalized as necessity. This seems very counterintuitive because it includes the release scheme $\mathbb{A}(A) \to A$, and as a result, in self-referential contexts we can prove that some statements formally derivable from the stated axioms are not assertible. (Let $L^\dagger$ be a sentence which states that its negation is assertible. Using release, we get $L^\dagger \to \neg L^\dagger$, which entails $\neg L^\dagger$. So we have proven $\neg L^\dagger$. But if $\neg L^\dagger$ were assertible, then we could infer $L^\dagger$ and get a contradiction. So we have proven a statement and also proven that it is not assertible.) [77] also discusses assertibility from a formal standpoint, but mainly in connection with the implication $A \to \mathbb{A}(A)$; a full axiomatization is not attempted.

30 (p. 116). The hard-nosed option of insisting that we actually do *not* have a proof that $2^{11} - 1$ is either prime or composite until we have determined which it is, is hardly viable. For instance, this position would render virtually all inductive reasoning illegitimate. We can infer that $2^{11} - 1$ is either prime or composite from the statement that every number is either prime or composite, which is easily proven using induction. So the appeal to induction would have to fail. (This example is taken from [68], p. 475, where it is used to argue for a distinction between "direct" and "indirect" proofs. The conclusion drawn

there is that a proof that there exists a "direct" proof of A is itself also a proof of A, but merely an "indirect" one. However, this suggestion runs into difficulty with the sentence "This sentence has no direct proof" because it should be immediate that every direct proof is sound, from which one can infer that the sentence indeed has no direct proof, which means that we can prove it but not directly.)

31 (p. 119). This solution is proposed, in exactly these terms, by Putnam ([72], p. 319). I hesitate to make a direct attribution because his definition of assertibility seems somewhat different from mine. But the two proposals are, if not literally identical, at least very close.

32 (p. 119). [58].

Chapter 5

1 (p. 123). Except for the material in Section 5.6, this is all (a revised version of) the contents of the series of papers [95] — [100].

2 (p. 130). This argument is made in Section 2 (d) of [28].

3 (p. 131). More technically, the formalized ω-consistency of PA is what matters.

4 (p. 132). [108].

5 (p. 132). These are mostly adapted from Section 3 of [108].

6 (p. 134). This formulation of the disjunctive trust problem was suggested to me by Cameron Freer.

7 (p. 136). I have not addressed Yudkowsky and Herreshoff's main concern, which involves a rational agent who is seeking not a license to perform a particular action, but general permission to delegate the performance of actions to a second agent. This is more difficult because the first agent does not merely have to sanction the second agent's judgement that some particular action has been licensed, as in the naturalistic trust paradox. Rather, the first agent is required to globally affirm the correctness of any such judgement the second agent might make. This leads to the familiar Löbian obstacle that the first agent is not even able to affirm the soundness of its own reasoning.

Bringing assertibility into the picture does not obviously help with this problem. However, it is unclear what kind of a solution one should hope for because granting the first agent the ability to delegate tasks to a second agent leads to a "procrastination paradox" wherein the possibility arises that the second agent might delegate those tasks to a third agent, and so on, indefinitely. So one would want to reformulate the problem in some way that will ensure delegating tasks to proxy agents would not mean that they might never be performed.

8 (p. 137). See [92].

9 (p. 148). [30].

10 (p. 149). [30].

Chapter 6

1 (p. 151). [47], p. 1.

2 (p. 151). [8], p. 615.

3 (p. 152). [84].

4 (p. 152). [84], p. 61.

5 (p. 152). [84], p. 62.

6 (p. 153). [64], p. 40.

7 (p. 153). [65], p. 506.

8 (p. 153). [67], p. 183.

9 (p. 153). [81], p. 637.

10 (p. 153). [82], pp. 322-323.
11 (p. 154). [54], p. 374.
12 (p. 154). Criticisms similar to this one appear in [65], p. 511 and [54], p. 374.
13 (p. 156). [78], p. 36. I learned of this comment from [10].
14 (p. 156). [20], p. 195.
15 (p. 156). [27], p. 441.
16 (p. 162). [59], p. 4.
17 (p. 162). [9], p. 491.
18 (p. 163). An interesting elaboration of this idea is given in [15].
19 (p. 164). [6].
20 (p. 165). This view is developed in [48].
21 (p. 171). [5].

Bibliography

[1] J. Barwise and J. Etchemendy, *The Liar: An Essay on Truth and Circularity*, 1987.

[2] J. C. Beall (ed.), *Revenge of the Liar*, 2007

[3] J. C. Beall and G. Restall, *Logical Pluralism*, 2005.

[4] M. J. Beeson, *Foundations of Constructive Mathematics*, 1985.

[5] J. S. Bell, On the Einstein-Podolsky-Rosen paradox, *Physics* **1** (1964), 195-200.

[6] P. Benacerraf, Mathematical truth, *J. Philos.* **70** (1983), 661-679; reprinted in [7], 403-420.

[7] P. Benacerraf and H. Putnam (eds.), *Philosophy of Mathematics: Selected Readings* (second edition), 1983.

[8] M. Black, The elusiveness of sets, *Review of Metaphysics* **24** (1971), 614-636.

[9] G. Boolos, The iterative conception of set, *J. Philos.* **68** (1971), 215-232; reprinted in [7], 486-502.

[10] R. T. Cook, Embracing revenge: on the indefinite extensibility of language, in [2], 31-52.

[11] D. Davidson, Truth and meaning, *Synthese* **17** (1967), 304-323; reprinted in [12], 17-42.

[12] ———, *Inquiries into Truth and Interpretation*, 1984.

[13] ———, The folly of trying to define truth, *J. Philos.* **93** (1996), 263-278.

[14] ———, *Truth & Predication*, 2005.

[15] E. B. Davies, Building infinite machines, *British J. Philos. Sci.* **52** (2001), 671-682.

[16] R. Dedekind, *Essays on the Theory of Numbers*, 1963.

[17] G. F. Díez, Five observations concerning the intended meaning of the intuitionistic logical constants, *J. Philos. Logic* **29** (2000), 409-424.

[18] M. Dummett, Truth, *Proc. Aristotelian Soc.* **59** (1959), 141-162; reprinted in [25], 1-24.

[19] ———, Wittgenstein's philosophy of mathematics, *Phil. Rev.* **68** (1959), 324-328; reprinted in [25], 166-185.

[20] ———, The philosophical significance of Gödel's theorem, *Ratio* **5** (1963), 140-155; reprinted in [25], 186-201.

[21] ———, *Frege: Philosophy of Language*, 1973.

[22] ———, The philosophical basis of intuitionistic logic, in *Logic Colloquium '73*, H. E. Rose and J. C. Shepherdson (eds.), 1975, 5-40; reprinted in [25], 215-247, and [7], 97-129.

[23] ———, What is a theory of meaning? (II), in *Truth and Meaning*, G. Evans and J. McDowell (eds.), 1976, 67-137; reprinted in [27], 34-93.

[24] ———, *Elements of Intuitionism*, 1977.

[25] ———, *Truth and Other Enigmas*, 1978.

[26] ———, Language and truth, in *Approaches to Language*, R. Harris (ed.), 1983, 95-125; reprinted in [27], 117-146.

[27] ———, *The Seas of Language*, 1993.

[28] S. Feferman, Transfinite recursive progressions of axiomatic theories, *J. Symb. Logic* **27** (1962), 259-316.

[29] H. Field, Deflationist views of meaning and content, *Mind* **103** (1994), 249-285.

[30] M. Forti and R. Hinnion, The consistency problem for positive comprehension principles, *J. Symbolic Logic* **54** (1989), 1401-1418.

[31] G. Frege, *Begriffsschrift, eine der arithmetischen nachgebildete Formelsprache des reinen Denkens* (1879); translated as *Begriffsschrift*, a formula language, modeled upon that of arithmetic, for pure thought, [93], 1-82.

[32] ———, Über Begriff und Gegenstand, *Vierteljahresschrift für wissenschaftliche Philosophie* **16** (1892), 192-205; translated as On concept and object, [34], 42-55.

[33] ———, Der Gedanke: Eine logische Untersuchung, *Beiträge zur Philosophie des deutschen Idealismus* **2** (1918-1919), 58-77; translated as The thought: a logical inquiry, *Mind* **65** (1956), 289-311.

[34] P. Geach and M. Black, *Translations from the Philosophical Writings of Gottlob Frege*, 1952.

[35] G. Gentzen, Untersuchungen über das logische Schliessen I, II, *Math. Z.* **39** (1935), 176-210 and 405-431.

[36] K. Gödel, Die Vollständigkeit der Axiome des logischen Funktionenkalküls, *Monatsh. Math. Phys.* **37** (1930), 349-360, translated as The completeness of the axioms of the functional calculus of logic, [93], 583-591.

[37] ———, Über formal unentscheidbare Sätze der Principia Mathematica und verwandter Systeme, I, *Monatsh. Math. Phys.* **38** (1931), 163-198, translated as On formally undecidable propositions of Principia Mathematica and related systems, I, [93], 595-616.

[38] ———, Eine Interpretation des intuitionistischen Aussagenkalküls, *Ergebnisse eines mathematischen Kolloquiums* **4** (1933), 39-40; translated as An interpretation of the intuitionistic propositional calculus, [41], 301.

[39] ———, What is Cantor's continuum problem?, *Amer. Math. Monthly* **54** (1947), 515-525.

[40] ———, Some basic theorems on the foundations of mathematics and their implications, in [42], 304-323.

[41] ———, *Collected Works, Vol. I* (S. Feferman et al., eds), 1986.

[42] ———, *Collected Works, Vol. III* (S. Feferman et al., eds), 1995.

[43] N. Goodman, A theory of constructions equivalent to arithmetic, in *Intuitionism and Proof Theory*, Kino, Myhill, and Vesley (eds.), 1970, 101-120.

[44] A. Gupta, A critique of deflationism, *Phil. Topics* **21** (1993), 57-81.

[45] A. Gupta and N. Belnap, *The Revision Theory of Truth*, 1993.

[46] V. Halbach, *Axiomatic Theories of Truth*, 2014.

[47] P. Halmos, *Naive Set Theory*, 1960.

[48] G. Hellman, *Mathematics without Numbers: Towards a Modal-Structural Interpretation*, 1989.

[49] L. Henkin, The completeness of the first-order functional calculus, *J. Symbolic Logic* **14** (1949), 159-166.

[50] A. Heyting, Sur la logique intuitionniste, *Acad. Roy. Belg. Bull. Cl. Sci.* **16** (1930), 957-963.

[51] ———, Die intuitionistische Grundlegung der Mathematik, *Erkenntnis* **2** (1931), 106-115; translated as The intuitionist foundations of mathematics, [7], 52-61.

[52] D. Hilbert, Über das Unendliche, *Math. Ann.* **95** (1926), 161-190, translated as On the infinite, [7], 183-201.

[53] P. Horwich, *Truth* (second edition), 1998.

[54] I. Jané, The iterative conception of sets from a Cantorian perspective, in *Logic, Methodology, and Philosophy of Science: Proceedings of the Twelfth International Congress*, P. Hájek, D. Westertål, and L. Valdés-Villanueva (eds.), 2005, 373-393.

[55] D. Jarden, A simple proof that a power of an irrational number to an irrational exponent may be rational, *Scripta Math.* **19** (1953), 229.

[56] S. C. Kleene, Recursive predicates and quantifiers, *Trans. Amer. Math. Soc.* **54** (1943), 41-73.

[57] G. Kreisel, Foundations of intuitionistic logic, in *Logic, Methodology and Philosophy of Science (Proc. 1960 Internat. Congr.)*, 1962, 198-210.

[58] S. Kripke, Outline of a theory of truth, *J. Philos.* **72** (1975), 690-716.

[59] P. Maddy, *Realism in Mathematics*, 1990.

[60] P. Martin-Löf, *Intuitionistic Type Theory* (notes by G. Sambin), 1984.

[61] ———, On the meanings of the logical constants and the justifications of the logical laws, *Nordic J. Philos. Logic* **1** (1996), 11-60.

[62] E. Martino, The priority of arithmetical truth over arithmetical provability, *Topoi* **21** (2002), 55-63.

[63] V. McGee, *Truth, Vagueness, and Paradox*, 1991.

[64] C. Menzel, On the iterative explanation of the paradoxes, *Philos. Stud.* **49** (1986), 37-61.

[65] C. Parsons, What is the iterative conception of set?, in *Logic, Foundations of Mathematics and Computability Theory, Part I*, 1977, 335-367; reprinted in [7], 503-529.

[66] I. Peano, *Arithmetices Principia, Novo Methodo Exposita*, 1889, translated as The principles of arithmetic, presented by a new method, [93], 83-97.

[67] M. D. Potter, Iterative set theory, *Philos. Quart.* **43** (1993), 178-193.

[68] D. Prawitz, Some remarks on verificationistic theories of meaning, *Synthese* **73** (1987), 471-477.

[69] ———, Truth and objectivity from a verificationist point of view, in *Truth in Mathematics (Mussomeli, 1995)*, 1998, 41-51.

[70] G. Priest, J. C. Beall, and B. Amour-Garb (eds.), *The Law of Non-Contradiction*, 2007.

[71] H. Putnam, *Reason, Truth and History* (1981).

[72] ———, On truth, in *How Many Questions? Essays in Honour of Sidney Morgenbesser*, L. Cauman et al. (eds.), 1983, 35-56; reprinted in [73], 315-329.

[73] ———, *Words & Life*, 1994.

[74] W. V. O. Quine, *The Ways of Paradox and Other Essays*, 1966.

[75] ———, *Mathematical Logic* (revised edition), 1981.

[76] ———, *Philosophy of Logic* (second edition), 1986.

[77] G. Restall, Constructive logic, truth and warranted assertibility, *Philos. Quart.* **51** (2001), 474-483.

[78] B. Russell, On some difficulties in the theory of transfinite numbers and order types, *Proc. London Math. Soc.* **4** (1907), 29-53.

[79] R. Schantz (ed.), *What is Truth?*, 2002.

[80] S. Shapiro, Proof and truth: through thick and thin, *J. Philos.* **95** (1998), 493-521.

[81] M. F. Sharlow, Proper classes via the iterative conception of set, *J. Symbolic Logic* **52** (1987), 636-650.

[82] J. R. Shoenfield, Axioms of set theory, in *Handbook of Mathematical Logic*, J. Barwise (ed.), 1977, 321-344.

[83] B. H. Slater, Concept and object in Frege, *Minerva* **4** (2000), http://www.minerva.mic.ul.ie//vol4/index.html; reprinted in [85], 99-112.

[84] ———, Grammar and Sets, *Australas. J. Philos.* **84** (2006), 59-73.

[85] ———, *The De-Mathematisation of Logic*, 2007.

[86] S. Soames, *Understanding Truth*, 1999.

[87] A. Tarski, The concept of truth in formalized languages, in [89], 152-278.

[88] ———, The semantic conception of truth and the foundations of semantics, *Philos. and Phenomenol. Res.* **4** (1944), 341-376.

[89] ———, *Logic, Semantics, Metamathematics* (second edition), 1983.

[90] A. Tarski and R. L. Vaught, Arithmetical extensions of relational systems, *Compos. Math.* **13** (1956-1958), 81-102.

[91] A. S. Troelstra and D. van Dalen, *Constructivism in Mathematics, Vol. I*, 1988.

[92] A. S. Troelstra and H. Schwichtenberg, *Basic Proof Theory* (2000).

[93] J. van Heijenoort (ed.), *From Frege to Gödel: A Source Book in Mathematical Logic, 1879-1931*, 1967.

[94] P. van Inwagen, Why I don't understand substitutional quantification, *Philos. Stud.* **39** (1981), 281-285.

[95] N. Weaver, Constructive truth and circularity, arXiv:0905.1681.

[96] ———, Intuitionism and the liar paradox, *Ann. Pure Appl. Logic* **163** (2012), 1437-1445.

[97] ———, Kinds of concepts, arXiv:1112.6124.

[98] ———, The semantic conception of proof, arXiv:1112.6126.

[99] ———, Reasoning about constructive concepts, in *Infinity and Truth*, C. Chong, Q. Feng, T. A. Slaman, and W. H. Woodin (eds.), 2014, 187-198.

[100] ———, Paradoxes of rational agency and formal systems that verify their own soundness, arXiv:1312.3626.

[101] S. Weinstein, The intended interpretation of intuitionistic logic, *J. Philos. Logic* **12** (1983), 261-270.

[102] M. Williams, On some critics of deflationism, in [79], 146-158.

[103] L. Wittgenstein, *Tractatus Logico-Philosophicus*, 1922.

[104] ———, *Philosophical Investigations* (second edition), translated by G. E. M. Anscombe, 1958.

[105] W. H. Woodin, Large cardinal axioms and independence: the continuum problem revisited, *Math. Intelligencer* **16** (1994), 31-35.

[106] C. Wright, *Truth and Objectivity*, 1992.

[107] ———, Why Frege did not deserve his *granum salis*, *Grazer Philos. Stud.* **55** (1998), 239-263.

[108] E. Yudkowsky and M. Herreshoff, Tiling agents for self-modifying AI, and the Löbian obstacle (2013), http://intelligence.org/files/TilingAgents.pdf.

Notation Index

$\mathcal{A}$, 10

L, 14

0, 15, 147

$'$, 15, 147

$+$, 15

$\cdot$, 15

$\hat{n}$, 15

$\neg A$, 16, 18, 39, 64

$A \wedge B$, 16

$A \vee B$, 16, 17

$A \to B$, 16, 17

$\top$, 17

$\perp$, 18, 39, 64

τ, 19, 23, 39

$\leftrightarrow$, 20

$\mathbb{N}$, 22

$\forall$, 22, 23

$\exists$, 22, 23

$A[t]$, 23

$\equiv$, 26

$\tilde{\tau}$, 26

$C[[x]]$, 35

$\mathbb{L}$, 38

$C[[t]]$, 38

$R[[t_1, t_2]]$, 38

$\hat{A}$, 40

$\overline{A}$, 62

$\langle A \rangle$, 73

$\mathrm{Der}[x, y]$, 73

$\mathbb{D}[x]$, 73

$\mathrm{Der}_\mathrm{S}[x, y]$, 74

$\mathbb{D}_\mathrm{S}[x]$, 74

$\mathrm{Sub}[x, y, z]$, 74

$G[\hat{g}]$, 74

$\mathrm{Con}(\mathrm{S})$, 76

$\mathrm{NSub}[x, y, z]$, 78

$G^*[\hat{g}^*]$, 78

$\mathbb{T}[x]$, 79

$e[\langle t \rangle]$, 81

L', 89

$\mathbb{A}(A)$, 106

$\dot{\perp}$, 123, 149

$\dot{\wedge}$, 123, 149

$\dot{\vee}$, 123, 149

$\dot{\to}$, 123, 149

$\dot{\mathbb{A}}$, 123

$\dot{\neg}$, 124

$\mathbb{A}^2(A)$, 126

M, 132

$A \rightsquigarrow B$, 135

$\Gamma \Rightarrow C$, 137

$\breve{A}$, 139

$x \, \varepsilon \, y$, 143

ω, 147

$\dot{\forall}$, 149

$\dot{\exists}$, 149

Subject Index

P-acceptance/rejection, 5
$\text{Comp}(\text{PF}_{\mathcal{T}}) + \text{D}$, 148
$(=)$ law, 144
$(\forall)$ law, 117
$(\forall_{lo})$ law, 117
$(\rightarrow)$ law, 112
$(\rightarrow_{lo})$ law, 111
$(\wedge)$ law, 112
$(\wedge_{lo})$ law, 111
(C) law, 112
(C_{lo}) law, 111
(Comp) law, 144
(I) law, 112

abstract object, 164
arithmetic, 22, 70
arithmetic mod n, 26
assertibility, 89
assertibility rule, 135
assertible consistency of PA^{A}, 129
assertible liar pair, 124
assertible liar paradox, 89
assertible liar sentence, 89
assertible soundness of PA^{A}, 129
assertible soundness of intuitionistic
  logic, 113
assertoric variable, 149
atomic formula, 22, 38
atomic numerical sentence, 16
Attacker, 138
axiom of choice, 100

axiom scheme, 48
axioms, 61

Barwise, Jon, 56
Basic Law V, 173
Bell, J. S., 171
Belnap, Nuel, 55, 56
binary operation symbol, 38
Black, Max, 151
bound variable, 23

capture law, 106
capture scheme, 111, 112
categorical sentence, 55
CC, 145
CC$'$, 147
CC$_\forall$, 148
CC$'_\forall$, 148
Church-Turing thesis, 72
circumscription, 158
classical consistency, 66
classical derivability, 64
classical logic, 64
complete consistency, 126
completeness theorem, 68
comprehension scheme, 144
concept, 31
concept symbol, 38
conclusion introduction rules, 138
conclusion of a sequent, 137
concrete proxy, 165

conjunction, 16
constant symbol, 15, 38
constant term, 39
construction, 110
Constructive Concepts, 145
correspondence theory, 10
count noun, 152

decidability, 160
Defender, 138
definite concept, 157
derivability in a system, 70
derivation, 63
determinate truth, 57
dialetheism, 172
discharging assumptions, 63
disjunction, 16, 17
disjunctive trust, 134
disquotation, 13
double negation elimination, 64
Dummett's fallacy, 29
Dummett, Michael, 27–30, 156–158,
    166

elimination rules, 63
entailment, 17
equivalence, 20
Etchemendy, John, 56
ex falso quodlibet, 64
expression of a concept, 32
expression of a proposition, 7
expression of a relation, 33
extension of a concept, 151
extension of PA, 74
extensionality axiom, 144

falling under, 32
finitely decidable axiomatization, 73
finitely decidable property, 71
finitism, 99
first incompleteness theorem, 75
first-order variable, 58
for all, 23
for some, 23
formal system, 70
formalism, 90

formula, 22, 39
free substitution, 63
free variable, 23
Frege, Gottlob, 5, 6, 31, 38, 52, 55,
    146, 173

Gödel number, 73
Gödel sentence, 74
Gödel, Kurt, 73, 75, 77
Gentzen, Gerhard, 62, 137
grounded truth, 57
Gupta, Anil, 55, 56

Halmos, Paul, 151
having a property, 32
Herreshoff, Marcello, 132
Hilbert axioms, 69
Hilbert system, 69
Hilbert, David, 3, 69, 90
holding, 32
Horwich, Paul, 49

identity criterion for concepts, 32,
    120
identity criterion for propositions, 8,
    120
identity criterion for relations, 33,
    120
implication, 16, 17
incoming sequent, 138
increasing formula, 136
indefinite extensibility, 156
indirect assertion, 11
induction, 36
infinite conjunction, 11
internal law, 112
interpretation, 39, 41
interpreted language, 41
intersubstitutability axiom, 145
introduction rules, 63
intuitionism, 95
intuitionistic derivability, 64
intuitionistic logic, 64
iterative conception of sets, 153

Jarden, Dov, 100

Kripke, Saul, 57, 119, 120, 134, 169

Löb's theorem, 79
Löb, Martin, 78
Lagrange, Joseph-Louis, 33
law of assertible noncontradiction, 107
law of excluded middle, 1
liar paradox, 1
liar sentence, 1
logical connective, 16
logical operator, 111
logical tightrope, 109

mass noun, 152
McGee, Vann, 55
meaning of a sentence, 7
mention and use for concepts, 36
mention and use for propositions, 13
mereological sum, 152
minimal derivability, 64
minimal logic, 64
minimalism, 48
model, 39
model existence theorem, 67
modelled language, 39
multiset, 137

name, 123
natural deduction, 62
naturalistic trust, 132
negation, 16, 18
nominalism, 4, 164
nominalization of a predicate, 32
nominalization of a sentence, 7
nonconstructive existence proof, 100
nonstandard model, 75
nullary operation symbol, 38
numeral, 15
numerical sentence, 16
numerical term, 15

objectivity, 102
objectual quantification, 51
ontology, 103
operation symbol, 38

ordinary variable, 10
outgoing sequent, 138

PA, 70
PA$'$, 77
PA$''$, 78
PA*, 78
PA^{A}, 128
PA†, 86
PA^{T}, 79
PAR, 124
pathological sentence, 55
Peano arithmetic, 70
Peano axioms, 70
platonism, 99
positive set theory, 148
possible-worlds nominalism, 166
power set, 154
predicate (formal), 33
predicativism, 99
prefixing a formula, 139
premise introduction rules, 138
premise of a sequent, 137
product operation, 15
proof, 89
proof scheme, 50
property, 31
proposition, 4
proposition symbol, 38
Pure Assertibility Reasoning, 124

quantifier, 22
quasi-quotation, 12
Quine, W. V. O., 2, 12, 14, 37, 74, 75, 105

rational agent, 132
realism, 4, 164
reasoning under assumptions, 62
recursive axiomatization, 73
referent of a predicate, 32
referent of a sentence, 7
reflective trust, 133
reflectively coherent trust, 133
relation, 33
relation symbol, 38

release law, 106
release rule, 112
revenge problem, 54
revised truth, 55
Russell concept, 37
Russell predicate, 37
Russell set, 38
Russell's paradox, 37
Russell, Bertrand, 6, 37, 146, 155, 156

schematic assertion, 10
schematic variable, 10
second incompleteness theorem, 77
second-order logic, 58
second-order Peano arithmetic, 147
second-order variable, 58
self-reproduction, 155
sentence (formal), 23, 39
sequent, 137
sequent calculus, 137
sequent derivation, 138
situational truth, 56
Slater, B. H., 152
Soames, Scott, 57
soundness, 66, 75
soundness of classical logic, 65
strengthened liar sentence, 54
substitutional quantification, 51
successor operation, 15
sum operation, 15
surveyability, 159

T-scheme, 9
T-scheme for holding, 37
Tarski, Alfred, 9, 10, 13–15, 46, 76, 83–88, 119, 120, 168, 169
Tarskian catastrophe, 13
term, 22, 38
there exists, 23
totality, 162
translation, 40
trivial sequent, 137
truth function, 19, 39
truth predicate, 14
truth value, 19

unary operation symbol, 38
undefinability theorem, 46, 76
universal closure, 62
universal modus ponens, 69
universal quantification, 23
universal sentence, 9
use and mention for concepts, 36
use and mention for propositions, 13

vague concept, 157
vague truth, 55
variant assertible liar sentence, 124
variant Russell concept, 149

weak falsehood, 125
weak interpretation, 136
winning strategy, 140
Wittgenstein, Ludwig, 4, 55
Wright, Crispin, 107

Yudkowsky, Eliezer, 132

Printed in the United States
By Bookmasters